SPATIAL ERROR ANALYSIS

IEEE Press
445 Hoes Lane, P.O. Box 1331
Piscataway, NJ 08855-1331

Aerospace & Electronic Systems Society, *Sponsor*
AES Liaison to IEEE Press, Myron Greenbaum

Cover Design: William T. Donnelly, *WT Design*

Technical Reviewers
Dr. Myron Kayton, *Kayton Engineering Company*
Dr. James R. Huddle, *Litton, GCS*
Dr. Allan J. Brockstein, *Litton Guidance and Control Systems*
Jack R. Harris, *TRACOR*

SPATIAL ERROR ANALYSIS

A Unified Application-Oriented Treatment

David Y. Hsu
Litton Guidance and Control Systems

IEEE PRESS

Aerospace & Electronic Systems Society, *Sponsor*

The Institute of Electrical and Electronics Engineers, Inc., New York

This book and other books may be purchased at a discount
from the publisher when ordered in bulk quantities. Contact:

IEEE Press Marketing
Attn: Special Sales
Piscataway, NJ 08855-1331
Fax: (732) 981-9334

For more information about IEEE PRESS products,
visit the IEEE Home Page: http://www.ieee.org/

10 9 8 7 6 5 4 3 2 1

ISBN 0-7803-3453-1
IEEE Order Number PC5705

Library of Congress Cataloging-in-Publication Data

Hsu, David Y., 1945–
Spatial error analysis: a unified, application-oriented treatment / David Y. Hsu.
p. cm.
"Aerospace & Electronic Systems Society, sponsor."
Includes bibliographical references and index.
ISBN 0–7803–3453–1
1. Electric engineering—Mathematics. 2. Error analysis (Mathematics) I. IEEE Aerospace and Electronic Systems Society. II. Title.
TK153.H78 1998
621.3′01′51—DC21 98–12430
CIP

In Loving Memory of My Mother

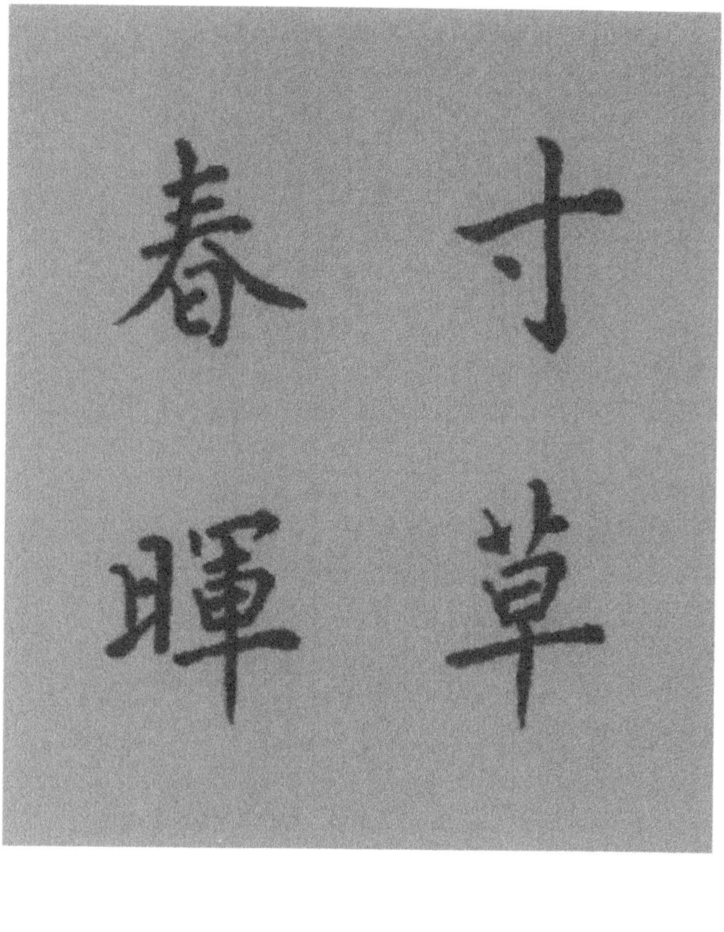

Contents

in this field will not need to go through the same struggling process as the author has had to do in the past.

This book is organized as follows:

Chapter 1 defines the notation to be used, classifies the type of problems to be investigated, and describes how to install the author-generated M-files for MATLAB.

Chapter 2 discusses two estimation methods of population parameters and quantifies the confidence level about interval estimation.

Chapter 3 treats error analysis for the one-dimensional problem. The relationships between root mean square (RMS) error, mean absolute error (MAE), and linear error probable (LEP) are explained.

Chapter 4 treats error analysis for the two-dimensional problem. The relationships between mean radial error (MRE), root mean square radial (RMSR) error, distance root mean square (DRMS) error, and circular error probable (CEP) are explained.

Chapter 5 treats error analysis for the three-dimensional problem. The relationships between mean radial error (MRE), root mean square radial (RMSR) error, and spherical error probable (SEP) are explained.

Chapter 6 shows how to find the maximum likelihood estimate (MLE) for the distribution of radial errors.

Chapter 7 discusses solutions for position location problems using error ellipses.

Chapter 8 defines the buyer's risk and seller's risk for a precision high-technology product and shows how a reasonable pass/fail criterion can be set so that it will be acceptable to both buyer and seller.

Appendices A, B, C provide a brief review of basic concepts in Statistics and Probability. These include (1) various probability density functions (PDFs) and cumulative density functions (CDFs), (2) method of confidence intervals, and (3) linear combination of random variables. They serve as a handy reference for those who have not had formal training in this area.

Appendix D shows the basic definitions of the various GPS dilution of precisions in terms of the time and position standard deviations.

Appendix E contains the listings for each of the basic author-generated M-files from which other, more involved, M-files are constructed. These programs represent one way to do the job; more clever and efficient approaches may exist. Suggestions for improvement from MATLAB users will be most welcome.

These M-file listings may also serve as a guide for the reader who uses computing environments other than MATLAB to develop a personal set of software tools.

The collection of M-files used in a chapter is summarized in a table toward the end of that chapter. The name of each M-file as well as its input(s) and output(s) are listed so that the reader will know exactly what to include and what to expect in the MATLAB environment.

I would like to express my appreciation to Dr. Allan J. Brockstein, research scientist at Litton Guidance and Control Systems, for his numerous valuable proofreading comments. Thanks are due to Dr. James R. Huddle, chief scientist at Litton Guidance and Control Systems, for his suggestion to include Chapter 8 in this book. I am indebted to Tao Wang from Personal TEX, Inc. for his timely technical support while I was learning PCTEX® to typeset the book. I would like to acknowledge the many helpful comments from Douglas M. Schwarz concerning the use of his Styled Text Toolbox in making special symbols for the graphs of this book. The Styled Text Toolbox is available to the public at the MathWorks FTP site.

I am also indebted to Linda Matarazzo of the IEEE Press for her smooth coordination of the editorial process.

Last, but not least, I am grateful to my father, General Si-Yen Hsu, for his love, trust, and constant encouragement. I also wish to thank my family—Charlotte, Henry, and Matthew—whose understanding made the completion of this book possible.

List of Figures

List of Tables

1

Introduction

This book serves as a source/reference book for engineers and scientists working with measurement errors in one-, two-, and three-dimensional space, as well as for people who desire to obtain a clear understanding of the concepts of the various error standards and their interrelationships.

Through MATLAB, this book also introduces and provides a convenient tool for computation and comparison so that table look-up and "guestimation" can be avoided.

Since the book is application oriented, only the important, relevant results from Probability and Statistics are used. For proofs of theorems and derivations, the reader can refer to excellent textbooks such as [**1**, **2**].

The reader is assumed to have had an introductory course in Probability and Statistics, and to have a fairly good working knowledge of Differential and Integral Calculus.

We will be concerned mainly with normally distributed random variables. The normal (Gaussian) distribution is useful because it seems to describe the random observations of most experiments. It also describes the distribution associated with the parameter estimation for most probability distributions.

The notation to be used throughout this book is defined in Section 1.1. All problems in error analysis can be cast into two categories: direct problems and inverse problems. In Sections 1.2 and 1.3 we shall define these two types of problems in order to set a pattern for discussion in the chapters to come. Section 1.4 shows how to use the author-generated programs to solve problems in navigation accuracy analysis.

1.1 NOTATION

We will write random variables symbolically in boldface type as, for example, the random variable **x**. Often, it is necessary to find the probability that the value of a random variable **x** is less than or equal to some real number x; this we write as

$$P\{\mathbf{x} \leq x\}$$

The notation shown in the following list will be used throughout this book, and any deviation from it will be noted immediately to avoid any confusion.

Let $\triangleq$ stand for "equal by definition."

- $[\ldots] \triangleq$ row vector
- $[\ldots]' \triangleq$ column vector
- $\vec{V}_1 = [x]' \triangleq$ position vector of one component
- $\vec{V}_2 = [x, y]' \triangleq$ position vector of two components
- $\vec{V}_3 = [x, y, z]' \triangleq$ position vector of three components
- $g(t) \triangleq$ probability density function of a random variable **t**
- $G(t) \triangleq$ cumulative distribution function of $g(t)$, $G(t) = \int_{-\infty}^{t} g(u)\,du$
- $P\{\vec{V} \in A\} \triangleq$ probability that the random vector $\vec{V}$ falls into region A
- $N(\mu, \sigma^2) \triangleq$ normal distribution with mean μ and variance σ^2

1.2 DIRECT PROBLEMS

Let the vectors $\vec{V}_1 = [\mathbf{x}]'$, $\vec{V}_2 = [\mathbf{x}, \mathbf{y}]'$, and $\vec{V}_3 = [\mathbf{x}, \mathbf{y}, \mathbf{z}]'$ represent one-, two-, and three-dimensional random vectors, with corresponding probability density functions, $f_1(x)$, $f_2(x, y)$, and $f_3(x, y, z)$, respectively.

The direct problem consists of finding the probability $p = P\{\vec{V} \in A\}$ when the region A is specified. Variables μ_x, μ_y, μ_z, and R are defined in Section 1.3.

For one-dimensional problems, A could be an interval specified as

$$|x - \mu_x| \leq R \tag{1.1}$$

For two-dimensional problems, A could be a circular region:

$$(x - \mu_x)^2 + (y - \mu_y)^2 \leq R^2 \tag{1.2}$$

or A could be an elliptical region with **elliptical scale** k

$$\frac{(x - \mu_x)^2}{a^2} + \frac{(y - \mu_y)^2}{b^2} \leq k^2 \tag{1.3}$$

For three-dimensional problems, A could be a spherical region:

$$(x - \mu_x)^2 + (y - \mu_y)^2 + (z - \mu_z)^2 \leq R^2 \tag{1.4}$$

or A could be an ellipsoidal region with **ellipsoidal scale** k

$$\frac{(x - \mu_x)^2}{a^2} + \frac{(y - \mu_y)^2}{b^2} + \frac{(z - \mu_z)^2}{c^2} \leq k^2 \tag{1.5}$$

Thus, the relationship between $f_1(x)$ and $p = P\{\vec{V}_1 \in A\}$ is

$$p = P\{\vec{V}_1 \in A\} = P\{x \in (\mu - R, \mu + R)\} = \int_{\mu-R}^{\mu+R} f_1(x)\, dx$$

Furthermore, the relationship between $f_2(x, y)$ and $p = P\{\vec{V}_2 \in A\}$ is

$$p = P\{\vec{V}_2 \in A\} = P\{(x, y) \in A\} = \iint_A f_2(x, y)\, dx\, dy$$

and the relationship between $f_3(x, y, z)$ and $p = P\{\vec{V}_3 \in A\}$ is

$$p = P\{\vec{V}_3 \in A\} = P\{(x, y, z) \in A\} = \iiint_A f_3(x, y, z)\, dx\, dy\, dz$$

The direct problem becomes "given the dimension of A, find $p = P\{\vec{V} \in A\}$," where the dimension of A is specified in terms of R or k in Equations (1.1) through (1.5).

If exact integration is possible, p can be expressed as a closed-form formula; otherwise p can be obtained through numerical integration of a single, double, or triple integral.

1.3 INVERSE PROBLEMS

For the inverse problem, the probability p is given, and one is to find, depending on dimensionality, the half-length R of the interval centered at the mean μ_x for the one-dimensional problem; the radius R or the scale k of the circle or ellipse centered at the mean (μ_x, μ_y) for the two-dimensional problem; and the radius R or the scale k of the sphere or ellipsoid centered at the mean (μ_x, μ_y, μ_z) for the three-dimensional problem, such that $p = P\{\vec{V} \in A\}$.

That is, given p, find the dimension of A in terms of R or k in Equations (1.1) through (1.5) such that

$$\int_{\mu-R}^{\mu+R} f_1(x)\, dx = p$$

or

$$\iint_A f_2(x, y)\, dx\, dy = p$$

or

$$\iiint_A f_3(x, y, z)\, dx\, dy\, dz = p$$

Two approaches are used in this book to solve the general inverse problem: given p, find R such that $F(R) = p$.

The first approach is via numerical trial-and-error. We make an initial guess R_0 of the true solution and compute $p_0 = F(R_0)$. If $p_0 > p$, decrease R_0; if $p_0 < p$, increase R_0. The process is repeated until adequate resolution is attained.

The second approach uses Newton–Raphson's method [3] to find the root of the equivalent problem

$$G(R) = F(R) - p = 0$$

We start with an initial guess R_0, and we iterate according to

$$R_{i+1} = R_i - G(R_i)/G'(R_i), \quad \text{for } i = 0, 1, 2, \ldots \tag{1.6}$$

where G' is the derivative of G, until a certain accuracy criterion is satisfied. This process is implemented in the M-file **newton.m**.

1.4 USE OF AUTHOR-GENERATED M-FILES

All the programs used in this book are MATLAB M-files. Each has been tested thoroughly with MATLAB Version 4.2c and Version 5.2 on various PC platforms (Pentium [166 MHz, 66 MHz], 486-60 MHz computers).

In order to install and use these M-files, the reader should follow these steps:

1. Create a subdirectory c:\mfile

   md c:\mfile
2. Copy all files from a:\mfile to c:\mfile

   copy a:\mfile*.* c:\mfile
3. (for Version 5.2) Start MATLAB.

   Click path browser to include c:\mfile in the MATLAB path.

   (for Version 4.2c) Include c:\mfile in the path of MATLAB by adding the line 'c:\mfile' in the file c:\matlab**matlabrc.m** through an editor. Start MATLAB.
4. Enter *nf2a*(1). The user should see the numerical result

   ```
   ans = 0.6827
   ```

   produced with an accompanying graph (see Figure 1.1) which describes the geometric meaning of the number 0.6827. This indicates a successful installation of the M-files.

The names and functions of these author-generated M-files are listed in the last section of each chapter. Those who wish to jump right in for a hands-on experience are encouraged to do so by referring to these sections in Chapters 1 through 8.

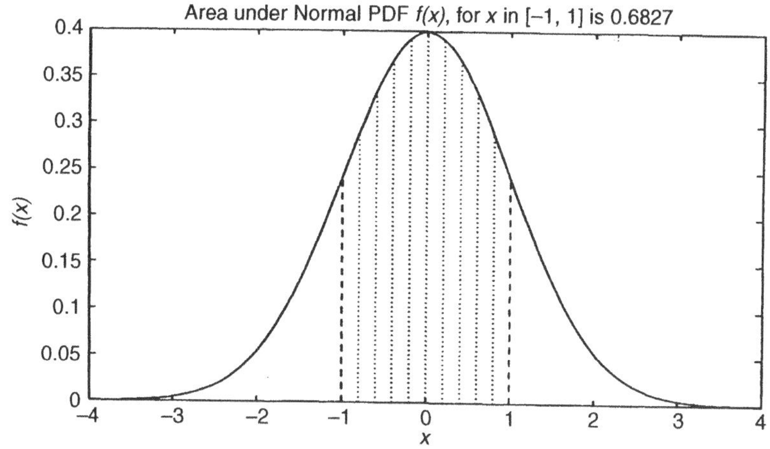

Figure 1.1 Results of the MATLAB Command $nf2a(1)$.

1.5 SUMMARY OF M-FILES

The M-files used or generated in this chapter are summarized in Table 1.1.

TABLE 1.1 M-files Used in Chapter 1

File Name	MATLAB Command
nf2a.m	$p = nf2a(r)$
nf1.m	$y = nf1(x)$
nf2.m	$p = nf2(r)$
pf3.m	$pf3('fname')$
newton.m	$x = newton(x_0,'fun','dfun', tol)$

The three files **nf1.m**, **nf2.m**, and **pf3.m** are called within the test program **nf2a.m**.

2

Parameter Estimation from Samples

All possible outcomes of a physical experiment constitute the **population space** of a random variable being observed; a finite set of observed outcomes of this experiment is called a **sample**.

The frequency with which an outcome occurs depends on the corresponding **probability density function (PDF)** associated with the population space.

Parameters of the population space such as mean and standard deviation are defined through the PDF. A random variable $\mathbf{t}$ with a PDF $f(t)$ will be denoted as $\mathbf{t} \sim f(t)$, and a random variable $\mathbf{z}$ with a standard normal probability density function

$$f(z) = \frac{1}{\sqrt{2\pi}} \exp\left[-\frac{z^2}{2}\right] \triangleq N(0, 1) \tag{2.1}$$

will be abbreviated as $\mathbf{z} \sim N(0, 1)$.

In general, if $\mathbf{x} \sim g(x)$, then its **mean value** or **expected value** is defined as

$$\mu_x = E[\mathbf{x}] = \int_{-\infty}^{\infty} x\, g(x)\, dx$$

and the **standard deviation** is defined as

$$\sigma_x = \sqrt{E[(\mathbf{x} - \mu_x)^2]} = \sqrt{\int_{-\infty}^{\infty} (x - \mu_x)^2\, g(x)\, dx}$$

The ratio of σ_x to μ_x expressed in units of percentage (%), $100\sigma_x/\mu_x$, is called the **coefficient of variation** of the random variable $\mathbf{x}$. We may drop the subscript and simply write μ or σ when the context makes it clear as to which random variable is being discussed. The term **variance** is used to refer to the square of the standard deviation or $var[\mathbf{x}] = \sigma_x^2$. Variance is also called the **second central moment** from the general definition of the kth central moment $E[(\mathbf{x} - \mu_x)^k]$.

Since it is sometimes impractical, if not impossible, to construct the entire population space, often a finite-sized sample is used to estimate the parameters of the population space.

In this book, Greek letters are used to denote population parameters; for example, μ is used for **population mean** and σ for **population standard deviation**. Roman letters are used to denote the corresponding estimates from a finite sample such as m for **sample mean** and s for **sample standard deviation**.

There are two types of estimation for the population parameters: **point estimation** and **interval estimation**. Both estimates are expressed in terms of the observed values of the random variable. Section 2.1 introduces point estimation for both population mean and population standard deviation. Section 2.2 presents two important theorems concerning sampling distributions. Sections 2.3 and 2.4 discuss the interval estimation of population mean and population standard deviation, respectively.

2.1 POINT ESTIMATE OF POPULATION PARAMETERS

When a set of N observations $\{x_1, x_2, \ldots, x_N\}$ of a random variable $\mathbf{x}$ is obtained from an experiment, the **sample mean** m and **sample standard deviation** s are computed via the point estimation formulas

$$m = \sum_{i=1}^{N} x_i/N \tag{2.2}$$

$$s = \sqrt{\sum_{i=1}^{N}(x_i - m)^2/(N-1)} \tag{2.3}$$

These two formulas indicate the dependence of the point estimates on the observations and are implemented in the built-in MATLAB M-files **mean.m** and **std.m**, respectively.

The term s^2 is also called an **unbiased estimate of the variance**. We will define "unbiased estimate" and show why s^2 is an unbiased estimate of σ^2 next[1].

A statistic y is called an **unbiased estimate** of the parameter η if the expected value of y equals η; that is, $E[y] = \eta$. Thus, to verify that s^2 is an unbiased estimate of σ^2, we need to show that

$$E[s^2] = \sigma^2$$

First, we know that if the random variable $\mathbf{x}$ is normally distributed with $N(\mu, \sigma^2)$ PDF, then the sample mean $\mathbf{m} = \sum_{i=1}^{n} x_i/N$ is also a normally distributed

[1]It can be shown that $E[s] = \{\Gamma(N/2)/\Gamma[(N-1)/2]\}\sqrt{2/(N-2)}\,\sigma \neq \sigma$, hence, s is a biased estimate of the standard deviation σ.

random variable with $N(\mu, \sigma^2/N)$ PDF. Now proceeding, we obtain

$$\begin{aligned}
E[s^2] &= E\left[\frac{1}{N-1}\sum_{i=1}^{N}(x_i - m)^2\right] \\
&= E\left\{\frac{1}{N-1}\sum_{i=1}^{N}[(x_i-\mu)-(m-\mu)]^2\right\} \\
&= E\left\{\frac{1}{N-1}\sum_{i=1}^{N}[(x_i-\mu)^2 - 2(x_i-\mu)(m-\mu) + (m-\mu)^2]\right\} \\
&= \frac{1}{N-1}E\left\{\sum_{i=1}^{N}[(x_i-\mu)^2 - 2(x_i-\mu)(m-\mu) + (m-\mu)^2]\right\} \\
&= \frac{1}{N-1}\left\{E\left[\sum_{i=1}^{N}(x_i-\mu)^2\right] - 2E\left[\sum_{i=1}^{N}(x_i-\mu)(m-\mu)\right] + E\left[\sum_{i=1}^{N}(m-\mu)^2\right]\right\} \\
&= \frac{1}{N-1}\left\{\sum_{i=1}^{N}E[(x_i-\mu)^2] - 2NE\left[\sum_{i=1}^{N}\frac{(x_i-\mu)}{N}(m-\mu)\right] + \sum_{i=1}^{N}E[(m-\mu)^2]\right\} \\
&= \frac{1}{N-1}\{N\sigma^2 - 2NE[(m-\mu)^2] + NE[(m-\mu)^2]\} \\
&= \frac{1}{N-1}\{N\sigma^2 - NE[(m-\mu)^2]\} \\
&= \frac{1}{N-1}\left\{N\sigma^2 - N\frac{\sigma^2}{N}\right\} \\
&= \frac{1}{N-1}[N\sigma^2 - \sigma^2] = \sigma^2
\end{aligned}$$

We have employed the fact

$$\sum_{i=1}^{N}\frac{(x_i-\mu)}{N} = m - \mu$$

to arrive at the conclusion.

An example will illustrate how these two files (mean.m, std.m) are used.

EXAMPLE 2.1

From the set of 100 observed values of a random variable **x** contained in the data file **eg2.1**, find its sample mean and sample standard deviation.

The four MATLAB commands

```
load c:\mfile\eg2.1

x = eg2

mean(x)

std(x)
```

show that $m = 0.1318$ and $s = 5.3140$.

Equation (2.3) shows that one needs to find the value of the sample mean m before s can be determined. The following equation provides a more straightforward approach:

$$s = \sqrt{\frac{N\sum_{i=1}^{N} x_i^2 - (\sum_{i=1}^{N} x_i)^2}{N(N-1)}}$$

The **mean deviation (MD)** for the sample $\{x_1, x_2, \ldots, x_N\}$ is defined as

$$\mathrm{MD} = \frac{\sum_{i=1}^{N} |x_i - m|}{N}$$

If we construct another sample $\{y_1, y_2, \ldots, y_N\}$ with $y_i = x_i + d$ for $i = 1, 2, \ldots, N$, d being a constant, then the two means differ by d. Nevertheless, the two mean deviations are identical, as seen in Example 2.2.

EXAMPLE 2.2

Given a sample of twelve elements

$$\{102, 115, 110, 109, 112, 121, 103, 113, 106, 114, 113, 123\}$$

find the mean and the mean deviation. Repeat the same process but first deduct 100 from each member of the sample.

Use the six MATLAB commands

```
x=[102, 115, 110, 109, 112, 121, 103, 113, 106, 114, 113, 123];
N = length(x);
xm1 = mean(x)
md1 = sum(abs(x-xm1))/N
xm2 = mean(x-100)
md2 = sum(abs(x-100-xm2))/N
```

to obtain $m_1 = 111.75$, $m_2 = 11.75$, and $\mathrm{MD}_1 = \mathrm{MD}_2 = 4.7917$.

In the CONTINUOUS case, the root mean square ($\mathrm{RMS} = \sqrt{E[\mathbf{x}^2]}$) and standard deviation (σ) of a random variable $\mathbf{x}$ with mean $\mu = E[\mathbf{x}]$ are related by

$$\begin{aligned}(\mathrm{RMS})^2 &= E[\mathbf{x}^2] = E[(\mathbf{x} - \mu + \mu)^2] \\ &= E[(\mathbf{x} - \mu)^2] + 2\mu E[(\mathbf{x} - \mu)] + \mu^2 \\ &= \sigma^2 + \mu^2 \end{aligned} \tag{2.4}$$

In the DISCRETE case, the RMS $(=\sqrt{\sum_{i=1}^{N} x^2/N})$ and the **biased estimate of the variance** $(s_b^2 = \sum_{i=1}^{N}(x_i - m)^2/N)$ of a random variable **x** with mean $m = \sum_{i=1}^{N} x_i/N$ are related by

$$\begin{aligned}(\mathrm{RMS})^2 &= \sum_{i=1}^{N} x_i^2/N = \sum_{i=1}^{N}(x_i - m + m)^2/N \\ &= \sum_{i=1}^{N}(x_i - m)^2/N + 2m\sum_{i=1}^{N}(x_i - m)/N + \sum_{i=1}^{N} m^2/N \\ &= \sum_{i=1}^{N}(x_i - m)^2/N + m^2 \\ &= s_b^2 + m^2 \qquad (2.5)\end{aligned}$$

In terms of the sample standard deviation s, we have

$$(\mathrm{RMS})^2 = \frac{N-1}{N}s^2 + m^2 \qquad (2.6)$$

If we use the dataset from Example 2.1 with the M-file **rmsx.m**, we will obtain RMS = 5.2890 via the command RMS = *rmsx(x)*.

2.2 SAMPLING DISTRIBUTION

The sample statistics defined in terms of the observations of a random variable are themselves random variables. Hence, one can talk about the **sampling distribution**, the PDF of the sample statistic.

At the beginning of Section 2.1, we defined the sample mean and sample standard deviation in Equations (2.1) and (2.2), respectively, as

$$m = \sum_{i=1}^{N} x_i/N \quad \text{and} \quad s = \sqrt{\sum_{i=1}^{N}(x_i - m)^2/(N-1)}$$

where $\{x_1, x_2, \ldots, x_N\}$ are N random samples of **x**. Note that nothing is said about the PDF of the random variable **x**.

For the discussion of the sampling distribution of these two statistics, we will assume that the random variable **x** is normally distributed.

We state without proof (see **[1, 4, 6]**) the following important results.

Theorem 2.2.1. *Suppose that* **x** *is a normal random variable with mean* μ *and variance* σ^2. *Let* $\mathrm{x}_1, \mathrm{x}_2, \ldots, \mathrm{x}_N$ *be* N *random observations of* **x**. *Then the random variable* **m** *has the normal distribution with mean* μ *and variance* σ^2/N, *and the random variable*

$$\frac{(N-1)\mathbf{s}^2}{\sigma^2} = \frac{1}{\sigma^2}\sum_{i=1}^{N}(x_i - \mathbf{m})^2$$

has the χ^2 *distribution with* (N − 1) *degrees of freedom. Furthermore,* **m** *and* $\mathbf{s}^2$ *are independent variables.*

Theorem 2.2.2. *Suppose that* $\mathbf{z}$ *is a standard normal random variable and* $\mathbf{v}$ *is a* χ^2 *random variable with* n *degrees of freedom, then if* $\mathbf{z}$ *and* $\mathbf{v}$ *are independent, the random variable*

$$\frac{\mathbf{z}}{\sqrt{\mathbf{v}/n}}$$

*has the student-*t *distribution with* n *degrees of freedom.*

Theorem 2.2.3. *If* $\mathbf{x}$ *is a normal random variable with mean* μ *and variance* σ^2*, then the random variable*

$$\frac{\mathbf{m} - \mu}{\mathbf{s}/\sqrt{N}}$$

*has the student-*t *distribution with* (N − 1) *degrees of freedom.*

The χ^2 and student-t distributions are discussed further in Appendix A.

2.3 INTERVAL ESTIMATE OF POPULATION MEAN

In the preceding section, point estimates of parameters have been considered. Often, however, one prefers an **interval estimate** that will express the accuracy of the estimate as well. This interval estimate provides a range into which the population parameter may fall. The end points of the interval are called the **confidence limits** for the parameter. Starting from a random sample of size N, we compute its mean m and standard deviation s. We then define an interval in terms of m and s, and determine whether the true population parameter can be located within this interval in a statistical sense.

The confidence associated with the estimate, or how close the estimate is to the true but unknown value of the population parameter, depends on the frequency or probability that the interval estimate actually captures the population parameter of concern. As expected, the confidence will grow with the number of random samples that are taken.

The accuracy associated with the confidence interval or the degree of confidence about this interval estimate is called the **confidence coefficient**. The degree of confidence we have for the population mean μ to be captured in the interval $(m - \varepsilon s, m + \varepsilon s)$ is expressed as the probability that the population parameter of interest falls into this interval centered at the sample mean m, and having a half-width of εs. Here s is the sample standard deviation and ε is a specified fraction, where

$$p = P\{\mu \in (m - \varepsilon s, m + \varepsilon s)\}$$

Note that p is dependent on both m and s from the sample. This equation can be rearranged to the following equivalent statements

$$\begin{aligned} p &= P\{m - \varepsilon s < \mu < m + \varepsilon s\} \\ &= P\{-\varepsilon < (m - \mu)/s < \varepsilon\} \\ &= P\{-\varepsilon\sqrt{N} < (m - \mu)\sqrt{N}/s < \varepsilon\sqrt{N}\} \end{aligned} \tag{2.7}$$

Since the random variable $(\mathbf{m} - \mu)\sqrt{N}/\mathbf{s}$ has a student-t distribution with $n = N - 1$ degrees of freedom, by Theorem 2.3, we have

$$\boldsymbol{\tau} = \mathbf{t}_{N-1} = (\mathbf{m} - \mu)\sqrt{N}/\mathbf{s} \quad \text{with PDF}$$

$$g(\tau) = \frac{\Gamma[(n+1)/2]}{\sqrt{n\pi}\Gamma(n/2)}(1 + \tau^2/n)^{-(n+2)/2}, \quad \text{for real } \tau \tag{2.8}$$

where $n = N - 1$, and the gamma function $\Gamma(x)$ is defined as

$$\Gamma(x) = \int_0^{\infty} t^{x-1} \exp^{-t} \, dt, \ \text{for } x > 0$$

Thus, the confidence coefficient p becomes

$$p = P\{-\varepsilon\sqrt{N} \leq \tau \leq \varepsilon\sqrt{N}\}$$

$$p = \int_{-\sqrt{N}\varepsilon}^{\sqrt{N}\varepsilon} \frac{\Gamma[(n+1)/2]}{\sqrt{n\pi}\Gamma(n/2)}(1 + \tau^2/n)^{-(n+2)/2} \, d\tau \tag{2.9}$$

When N is large, numerical evaluation of Equation (2.9) becomes troublesome due to growth of the gamma function. The integration can be carried out, for $N \geq 100$, with substitution of the standard normal distribution

$$p = \int_{-\sqrt{N}\varepsilon}^{\sqrt{N}\varepsilon} \frac{1}{\sqrt{2\pi}} \exp[-x^2/2] \, dx$$

The M-file **mucnf.m** is generated to compute the confidence coefficient of the interval estimate for the population mean when various interval lengths (expressed in terms of ε) and various sample sizes N are given.

EXAMPLE 2.3

How certain can we feel about the true population mean μ actually being within the interval $(m - \varepsilon s, m + \varepsilon s)$, with 60 observed sample values of a random variable **x**, when $\varepsilon = 25\%$?

The single MATLAB command

```
p = mucnf(0.25, 60)
```

gives the confidence coefficient $p = 94.24\%$.

Figure 2.1 shows the confidence coefficient corresponding to the interval estimation of μ for sample size N ranging from 1 to 1000 with $\varepsilon = 0.01$, 0.02, ..., 0.05, 0.1, 0.2, 0.5. Note that when $N = 1$, the confidence coefficient is 0.

Figure 2.2 contains the same information as in Figure 2.1, except that a semilog scale is employed to show clearer relationships when the sample size is small.

Figure 2.3 displays the standard normal distribution and student-t_n distribution $g(\tau)$ of Equation (2.8) for $n = 100$. Here it is seen that the two distribution

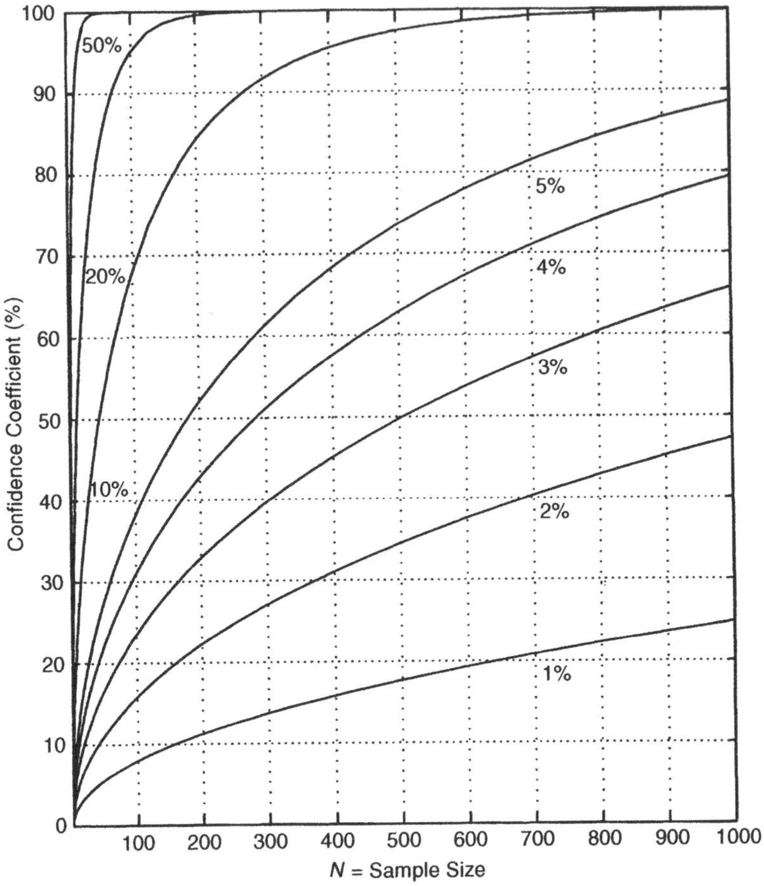

Figure 2.1 Confidence Coefficient for Interval Estimation of Population Mean.

functions practically coincide for large n, justifying replacement of the student-t_n PDF with the normal PDF in obtaining p through numerical integration.

2.4 INTERVAL ESTIMATE OF POPULATION STANDARD DEVIATION

The degree of confidence we have for the interval estimation of the population standard deviation σ is expressed as the probability that the population standard deviation σ falls into this interval centered at the sample standard deviation s, and having a half-width of εs, where ε is a specified fraction; that is,

$$p = P\{\sigma \in (s(1 - \varepsilon), s(1 + \varepsilon))\}$$

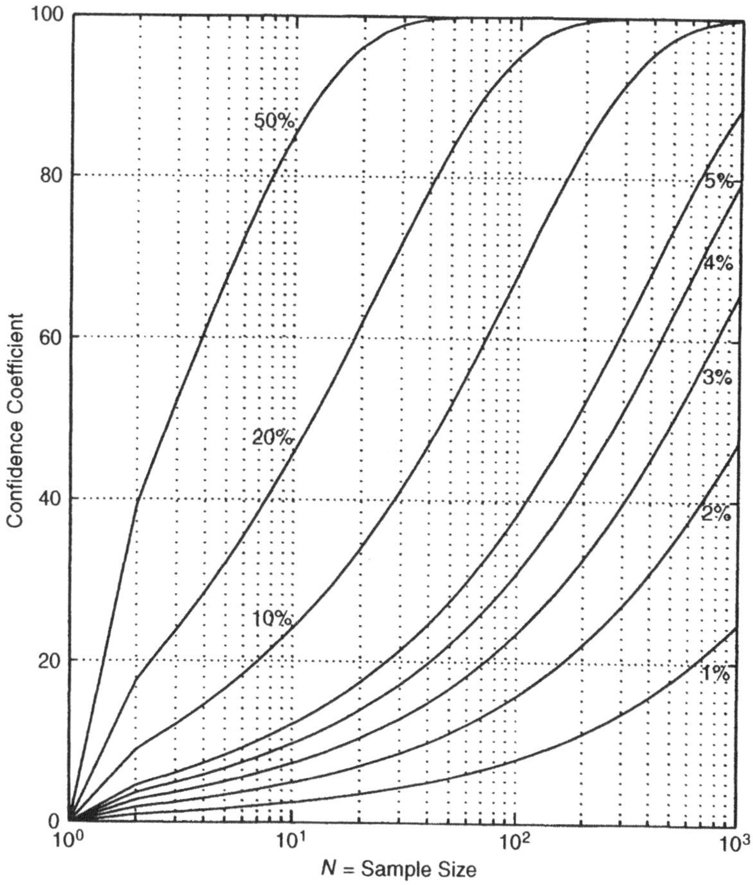

Figure 2.2 Confidence Coefficient for Interval Estimation of Population Mean on a Semilog Plot.

Note that p is dependent only on s from the sample. The preceding equation can be transformed to the following equivalent statements:

$$\begin{aligned} p &= P\{(1-\varepsilon) < \sigma/s < (1+\varepsilon)\} \\ &= P\{1/(1-\varepsilon) > s/\sigma > 1/(1+\varepsilon)\} \\ &= P\{(N-1)/(1-\varepsilon)^2 > s^2(N-1)/\sigma^2 > (N-1)/(1+\varepsilon)^2\} \end{aligned} \tag{2.10}$$

Since the random variable $\mathbf{s}^2(N-1)/\sigma^2$ has a χ^2 distribution with $n = N-1$ degrees of freedom, by Theorem 2.1,

$$\boldsymbol{\tau} = \boldsymbol{\chi}^2_{N-1} = \mathbf{s}^2(N-1)/\sigma^2 \quad \text{with PDF}$$

$$h(\tau) = \frac{1}{2^{(n/2)}\Gamma(n/2)} \tau^{\frac{n-2}{2}} \exp(-\tau/2) \tag{2.11}$$

where $n = N-1$.

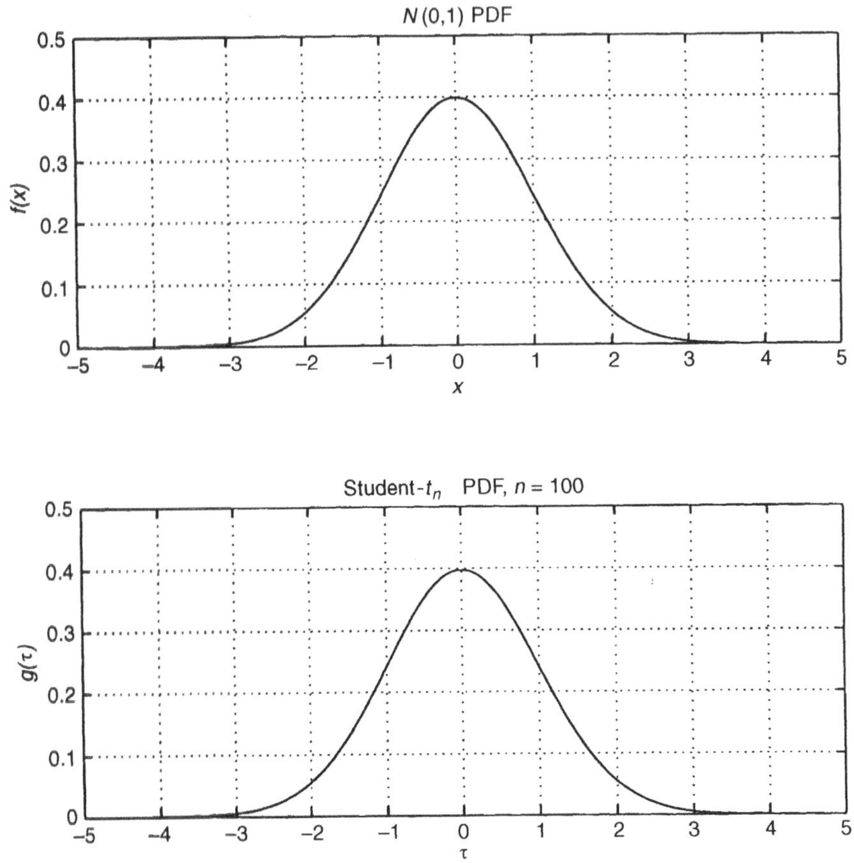

Figure 2.3 Comparison of the Standard Normal PDF with the Student-t_n PDF for $n = 100$.

Thus, the confidence coefficient p becomes

$$p = P\{(N-1)/(1+\varepsilon)^2 < \tau < (N-1)/(1-\varepsilon)^2\}$$

$$p = \int_{(N-1)/(1+\varepsilon)^2}^{(N-1)/(1-\varepsilon)^2} \frac{1}{2^{(\frac{N-1}{2})}\Gamma[(N-1)/2]} \tau^{\frac{N-3}{2}} \exp(-\tau/2)\, d\tau \tag{2.12}$$

As the sample size N increases, the gamma function $\Gamma[(N-1)/2]$ grows beyond machine limits and the numerical evaluation of Equation (2.12) becomes inaccurate. However, in this situation, we turn to the relationship that

$$\mathbf{u} = \sqrt{2\chi_n^2} - \sqrt{2n-1} = \sqrt{2\boldsymbol{\tau}} - \sqrt{2n-1} \sim N(0,1)$$

and transform the confidence coefficient as follows:

$$
\begin{aligned}
p &= P\{(N-1)/(1-\varepsilon)^2 > s^2(N-1)/\sigma^2 > (N-1)/(1+\varepsilon)\} \\
&= P\{(N-1)/(1-\varepsilon)^2 > \chi^2_{N-1} > (N-1)/(1+\varepsilon)\} \\
&= P\{\sqrt{2(N-1)}/(1-\varepsilon) > \sqrt{2\chi^2_{N-1}} > \sqrt{2(N-1)}/(1+\varepsilon)\} \\
&= P\left\{\frac{\sqrt{2n}}{(1-\varepsilon)} - \sqrt{2n-1} > \sqrt{2\chi^2_n} - \sqrt{2n-1} > \frac{\sqrt{2n}}{(1+\varepsilon)} - \sqrt{2n-1}\right\}
\end{aligned} \tag{2.13}
$$

Since $\boldsymbol{\tau} = \boldsymbol{\chi}^2_n \sim h(\tau)$ and $\mathbf{u} \approx \sqrt{2\boldsymbol{\tau}} - \sqrt{2n-1}$, therefore, $\boldsymbol{\tau} \approx (\mathbf{u} + \sqrt{2n-1})^2/2$ and the PDF $g(u)$ of $\mathbf{u}$ is obtained from

$$
\begin{aligned}
g(u) &\approx h(\tau)\frac{d\tau}{du} \\
&= h\left[\frac{(u+\sqrt{2n-1})^2}{2}\right](u+\sqrt{2n-1}), \qquad \text{for } u \geq -\sqrt{2n-1} \\
&= \frac{(u+\sqrt{2n-1})^{n-1}}{2^{(n-1)}\Gamma(n/2)}\exp\left[-\frac{(u+\sqrt{2n-1})^2}{4}\right]
\end{aligned} \tag{2.14}
$$

which can be shown to be a very close approximation to $N(0, 1)$, the PDF of a standard normal distribution. This leads to

$$
p = \int_{\sqrt{2n}/(1+\varepsilon)-\sqrt{2n-1}}^{\sqrt{2n}/(1-\varepsilon)-\sqrt{2n-1}} \frac{1}{\sqrt{2\pi}}\exp\left[-\frac{u^2}{2}\right]du \tag{2.15}
$$

The M-file **cnf.m** is generated to compute the confidence coefficient for the interval estimate of the population standard deviation given various interval lengths (expressed in terms of ε) and various sample sizes N.

EXAMPLE 2.4

How confident can we feel about the interval estimate $((1-\varepsilon)s, (1+\varepsilon)s)$ actually capturing the true population standard deviation σ when the sample size is 30 and $\varepsilon = 15\%$?

The single MATLAB command

```
p = cnf(0.15, 30)
```

gives the confidence coefficient $p = 74.15\%$.

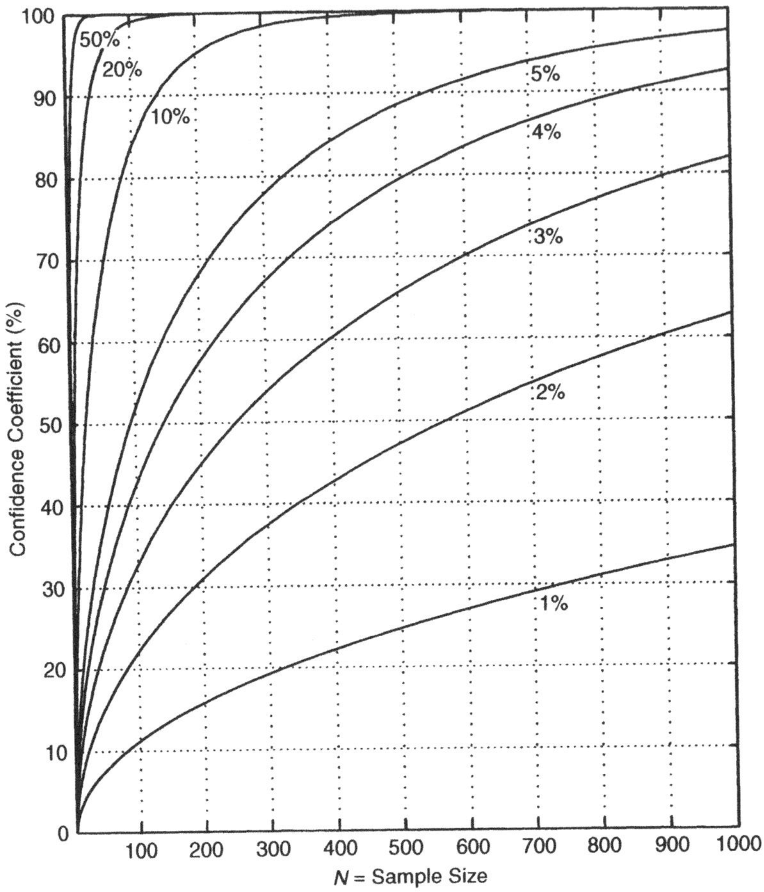

Figure 2.4 Confidence Coefficient for Interval Estimate of Population Standard Deviation.

Figure 2.4 shows the confidence coefficient corresponding to the interval estimate of σ using samples of size N ranging from 1 to 1000 for various ε values.

Figure 2.5 contains the same information as in Figure 2.4 except that a semilog scale is employed to show the clearer relationships when the sample size is small.

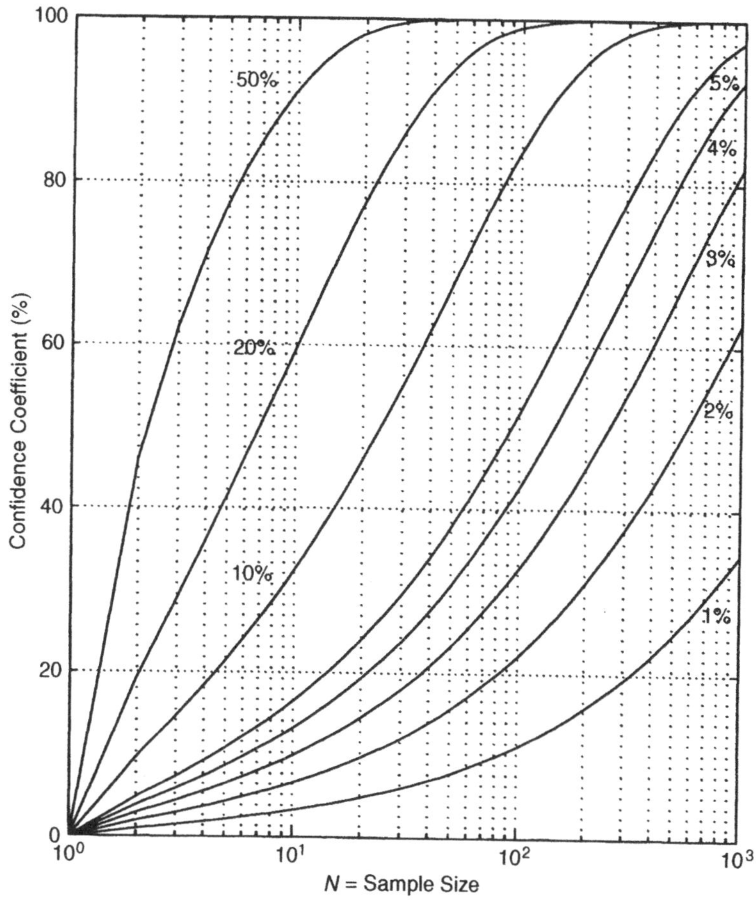

Figure 2.5 Confidence Coefficient for Interval Estimate of Population Standard Deviation on a Semilog Plot.

Figure 2.6 displays the standard normal distribution $f(x)$ of Equation (2.1) and the PDF $g(u)$ of Equation (2.14) for $n = 100$. Note that this function $g(u)$ is defined only for $u \geq -\sqrt{2n-1}$ or $u \geq -\sqrt{199} = -14.11$, when $n = 100$. The two curves practically coincide, justifying the computation of p using $N(0, 1)$ instead of the χ_n^2 distribution.

2.5 SUMMARY OF M-FILES

The M-files used in this chapter are summarized in Table 2.1

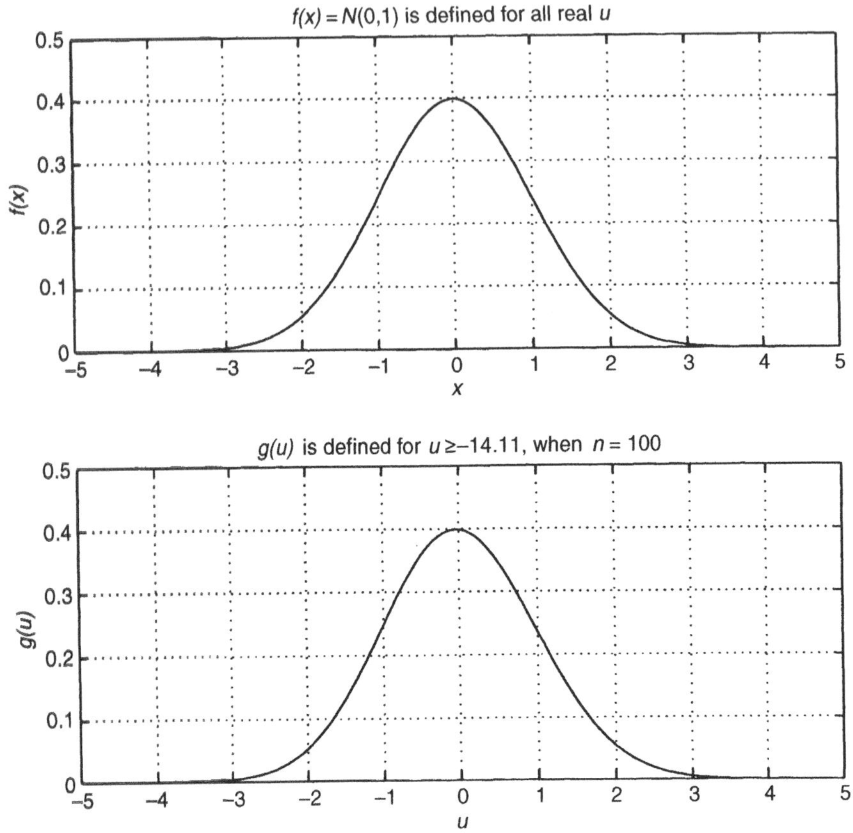

Figure 2.6 Comparison of the Standard Normal PDF with the PDF $g(u)$ for $n = 100$.

TABLE 2.1 M-files Used in Chapter 2

File Name	MATLAB Command
rmsx.m	$y = rmsx(x)$
nf1.m	$y = nf1(x_0)$
tandn.m	*tandn*
xandn.m	*xandn*
tdis.m	$y = tdis(x)$
gofu.m	$y = gofu(u)$
x2df.m	$y = x2df(x)$
x2dg.m	$y = x2dg(x)$
x2cdf.m	$p = x2cdf(x)$
cnf.m	$r = cnf(\varepsilon, n)$
mucnf.m	$p = mucnf(\varepsilon, n)$

The M-files **tandn.m** and **xandn.m** are used to generate Figures 2.3 and 2.6, respectively.

The standard normal distribution is implemented in the M-file **nf1.m**.

The student-t_n distribution with n degrees of freedom (see Equation [2.8]) is implemented in the M-file **tdis.m**. It is invoked in the program tandn.m.

The PDF $g(u)$ of the random variable $\mathbf{u}$ (see Equation [2.14]) is implemented in the M-file **gofu.m**. It is invoked in the program xandn.m.

The PDF and CDF of the χ^2 distribution with n degrees of freedom are implemented in the M-files **x2df.m** and **x2cdf.m**, respectively.

The file **x2dg.m** is needed for **x2cdf.m**, according to Equation (A.3) in Appendix A.

3

One-Dimensional Error Analysis

We start error analysis with the simplest kind—one-dimensional error analysis. Taking measurements of the height of a person, for example, or any measurement with a single degree of freedom, gives rise to this type of problem.

We will be concerned mainly with normally distributed random variables. The normal (Gaussian) distribution is useful because it describes the random observations of most experiments, and it describes well the distribution of the estimation of parameters from most probability models (via the central limit theorem; see Appendix A).

3.1 NORMAL DISTRIBUTION

The probability density function for one-dimensional errors, treated as a random variable **t**, is generally taken to be a normal distribution

$$g(t, \mu, \sigma) = \frac{1}{\sqrt{2\pi}\sigma} \cdot \exp\left[\frac{-(t-\mu)^2}{2\sigma^2}\right], \quad \text{for real } t$$

where μ is the mean and σ is the standard deviation of the random variable **t**. A short-hand notation for this is $\mathbf{t} \sim N(\mu, \sigma^2)$.

The curves in the upper graph of Figure 3.1 show how the peak location of the PDF shifts as the mean value varies for $\sigma = 1$. The curves in the lower graph show how the dispersion (or scatter) changes as the standard deviation value varies for $\mu = 0$.

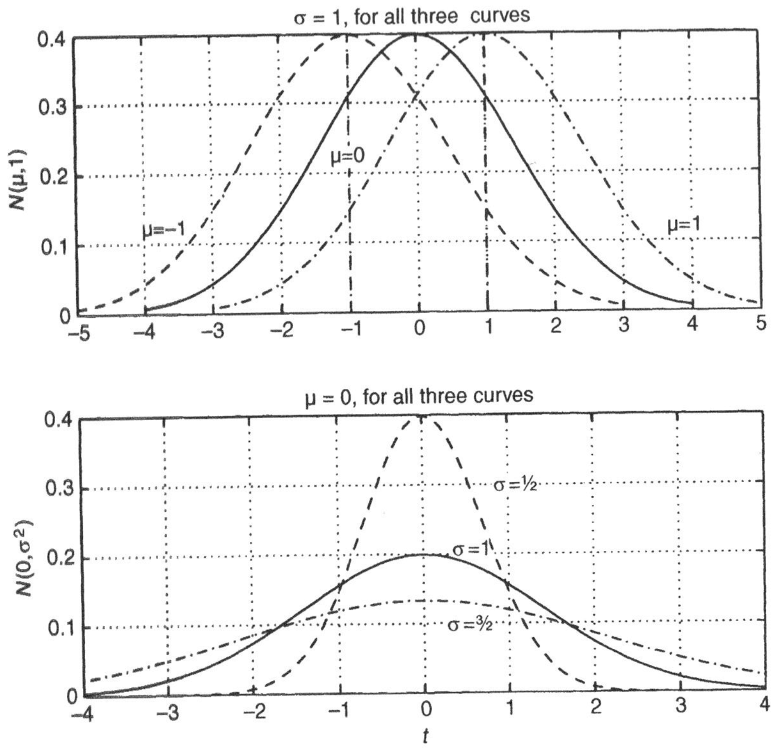

Figure 3.1 Normal Distribution.

After the mean μ is removed, the new random variable $\boldsymbol{\tau} = \mathbf{t} - \mu$ has the probability density function (PDF)

$$h(\tau, \sigma) = \frac{1}{\sqrt{2\pi}\sigma} \cdot \exp\left[\frac{-\tau^2}{2\sigma^2}\right], \quad \text{for real } \tau \tag{3.1}$$

This new random variable $\boldsymbol{\tau}$ is said to be normally distributed with zero-mean. Most random errors encountered in applications are of this type.

A further scaling of the random variable $\boldsymbol{\tau}$ by the standard deviation σ will lead to a new variable $\mathbf{x} = \boldsymbol{\tau}/\sigma$ with the probability density function

$$f(x) = \frac{1}{\sqrt{2\pi}} \cdot \exp\left[\frac{-x^2}{2}\right], \quad \text{for real } x \tag{3.2}$$

The random variable $\mathbf{x}$ is said to have a **standard normal distribution** and is commonly abbreviated with the notation $\mathbf{x} \sim N(0, 1)$. Note that $f(x)$ is an even function in x and $h(\tau, \sigma)$ is an even function in τ.

The standard normal distribution will be used throughout this book since any other normally distributed random variable $\mathbf{t} \sim N(\mu, \sigma^2)$ can be transformed to $\mathbf{x} \sim N(0, 1)$ via the formula (see [5]):

$$\mathbf{x} = \frac{\mathbf{t} - \mu}{\sigma} \tag{3.3}$$

This transformation first removes the mean μ and then normalizes the difference $\boldsymbol{\tau} = \mathbf{t} - \mu$ with respect to the standard deviation σ.

Figure 3.2 shows a bell-shaped normal PDF with $\mu = 0$ and $\sigma = 1$. This PDF has a peak value of $1/\sqrt{2\pi} \approx 0.3989$ at $x = 0$ and slowly decreases toward zero for $|x|$ greater than 4.

The probability that the random variable $\mathbf{t}$ falls in the interval $[-R + \mu, R + \mu]$ is equivalent to the probability of finding the random variable $\boldsymbol{\tau}$ in the interval $[-R, R]$, and is also equivalent to the probability that the random variable $\mathbf{x}$ falls in the interval $[-R/\sigma, R/\sigma]$. This probability is found from

$$\begin{aligned} p &= P\{\mu - R \leq \mathbf{t} \leq \mu + R\} \\ &= P\{-R \leq \tau \leq R\} = \int_{-R}^{R} h(\tau, \sigma)\, d\tau = 2\int_{0}^{R} h(\tau, \sigma)\, d\tau \\ &= P\{-r \leq \mathbf{x} \leq r\} = \int_{-r}^{r} f(x)\, dx = 2\int_{0}^{r} f(x)\, dx \end{aligned} \tag{3.4}$$

where r is the normalized half-interval length, $r = R/\sigma$ (with R being the half-interval length).

The preceding equation shows that the probability p is a function of the ratio (R/σ) only. We will adopt the notation $p = H(r) = H(R/\sigma)$ to associate with the PDF $h(r, \sigma)$ in Equation (3.1) for the random variable $\mathbf{r}$. This function, $H(r) = H(R/\sigma) = 2\int_0^R h(\tau, \sigma)\, d\tau$, is called the **cumulative distribution function** (or **CDF**) of the random variable $\mathbf{r}$. Thus, when working with CDF, we use only one variable, (R/σ) or r. When working with PDF, we need to use both R and σ.

Geometrically, this probability represents the area under the PDF curve $h(\tau, \sigma)$ above the τ-axis and bounded by vertical lines at $\tau = -R$ and $\tau = R$ (or the area under

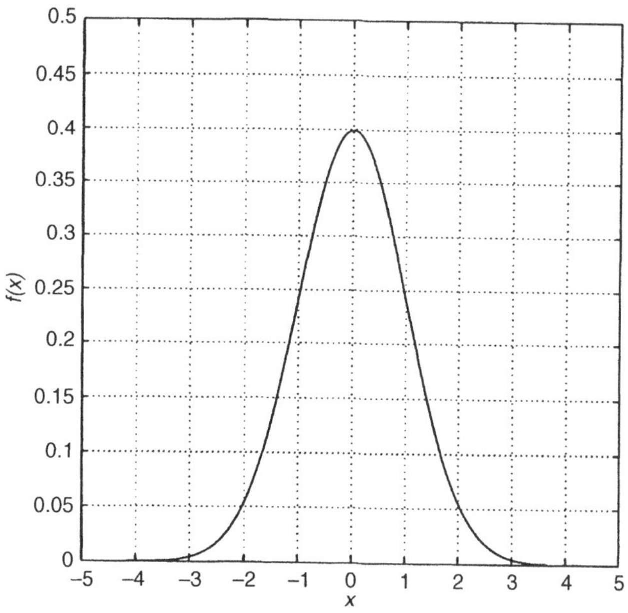

Figure 3.2 PDF for the Normal Distribution.

the PDF curve $f(x)$ above the x-axis and bounded by vertical lines at $x = -r$ and $x = r$).

3.2 ONE-DIMENSIONAL ERROR MEASURES

Standard error, 1-sigma error, RMS error, mean absolute error, and linear error probable (LEP or probable error—PE) will be defined in this section. We will use the random variable **t** with an $N(\mu, \sigma^2)$ distribution or

$$g(t, \mu, \sigma) = \frac{1}{\sqrt{2\pi}\sigma} \cdot \exp\left[\frac{-(t-\mu)^2}{2\sigma^2}\right], \quad \text{for real } t$$

to carry out our discussion.

The notation $E[w(\mathbf{t})]$ means the expected value of the random variable $w(\mathbf{t})$; it is expressed as

$$E[w(\mathbf{t})] = \int_{-\infty}^{\infty} w(t) g(t, \mu, \sigma)\, dt$$

from which we see $E[\mathbf{t}] = \int_{-\infty}^{\infty} t g(t, \mu, \sigma)\, dt = \mu$.

3.2.1 Standard Error

The standard error is defined to be the square root of the variance, or

$$\sqrt{E[(\mathbf{t}-\mu)^2]} = \sigma$$

The terms **standard error, standard deviation**, and **one-sigma (1σ) error** all have the same meaning in reference to this type of one-dimensional error.

3.2.2 Root Mean Square (RMS) Error

The root mean square (RMS) error equals the standard error only when the mean (μ) is zero. Starting from the definition of RMS

$$\text{RMS} = \sqrt{E[\mathbf{t}^2]} = \sqrt{\int_{-\infty}^{\infty} t^2 g(t, \mu, \sigma)\, dt} = \sqrt{\sigma^2 + \mu^2}$$

we can see that the RMS equals the standard error (or 1σ), when the random variable has zero mean ($\mu = 0$).

3.2.3 Mean Absolute Error (MAE)

Another error of interest is the mean absolute error (MAE), which is defined to be

$$\text{MAE} = E[|\mathbf{t}|] = \int_{-\infty}^{\infty} |t| g(t, \mu, \sigma)\, dt = \sqrt{(2/\pi)}\sigma \approx 0.7979\sigma$$

3.2.4 Linear Error Probable (LEP)

The **linear error probable (LEP),** or linear probable error, is defined to be the half-width of the interval centered at the mean, so that the area over this interval under the PDF is $\frac{1}{2}$. In terms of $\mathbf{t} \sim N(\mu, \sigma^2)$, LEP is solved numerically from

$$p = 0.5 = \int_{\mu-\text{LEP}}^{\mu+\text{LEP}} g(t, \mu, \sigma)\, dt$$

and LEP $= 0.6745\sigma \approx 2\sigma/3$. Sometimes LEP is referred to as **probable error (PE).**

Figure 3.3 shows both the PDF $h(r, \sigma)$ and CDF $H(r)$ of a standard normal random variable $\mathbf{r}$ for $r \geq 0$. The abscissa of the point marked with a "o" corresponds to the linear error probable (LEP, $r \approx 0.6745$), indicating that 50% of the sample points can be expected to lie within the interval $[-0.6745, 0.6745]$. The abscissa of the point marked with a "+" corresponds to 1 MAE ($r \approx 0.7979$), and its ordinate is approximately equal to 57.51%. The point marked with a "*" has a coordinate pair of $(1, 0.6827)$. This means that 68.27% of the sample points fall between $[-1, 1]$; or $p = 68.27\%$ is the 1σ probability. Other points of interest are the 2σ and 3σ probabilities, which are 95.45% and 99.73%, respectively.

Since the variable r in $H(r)$ is to be interpreted as a normalized quantity ($r = R/\sigma$), the solid CDF curve in Figure 3.3 is valid for all positive values of σ. The

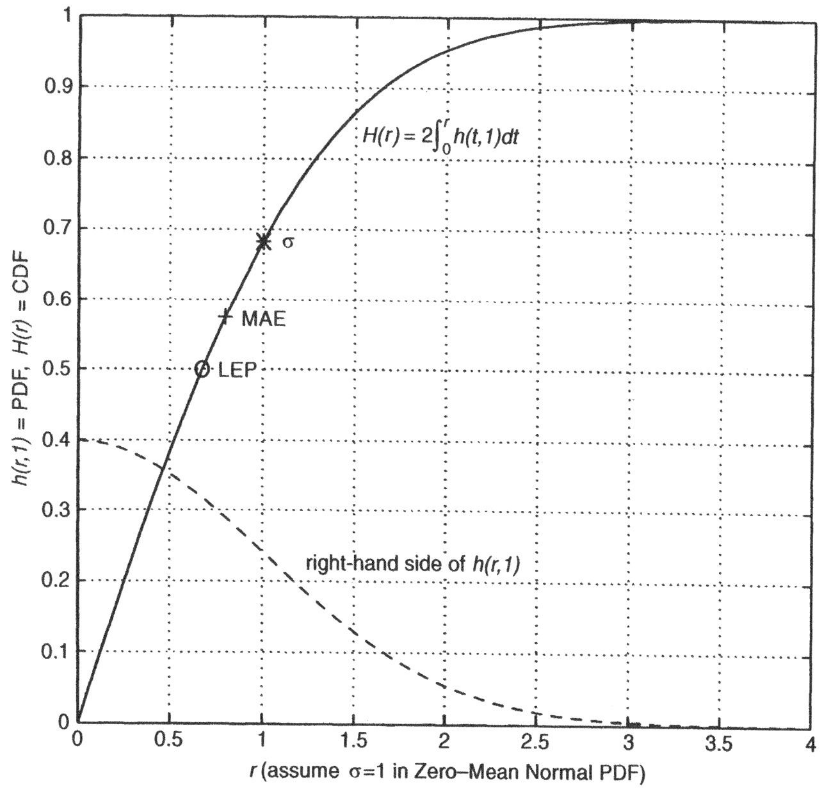

Figure 3.3 PDF and CDF for the Normal Distribution.

variable r in $h(r, \sigma)$ should not be treated as a relative quantity. Thus, the dashed PDF curve in Figure 3.3 is valid only for $\sigma = 1$. In other words, the variable used for the horizontal axis has two interpretations, one for the CDF and another for the PDF.

3.3 DIRECT PROBLEMS

The goal of a direct problem is to find the probability that the observed error falls in a given error bound, that is, to find p from R or $r = R/\sigma$, such that

$$p = H(R/\sigma) = \int_{-R}^{R} h(\tau, \sigma)\, d\tau \tag{3.5}$$

or

$$p = H(r) = \int_{-r}^{r} h(\tau, 1) d\tau = 2 \int_{0}^{r} \frac{1}{\sqrt{2\pi}} \exp\left[\frac{-\tau^2}{2}\right] d\tau \tag{3.6}$$

Since the error function

$$\operatorname{erf}(z) = (2/\sqrt{\pi}) \int_{0}^{z} \exp(-v^2)\, dv$$

is a built-in function, **erf.m** in MATLAB, the direct problem can be easily solved. Substituting $\tau = \sqrt{2}\, v$ into Equation (3.6), we obtain

$$p = H(r) = H(R/\sigma) = \operatorname{erf}(r/\sqrt{2}) \tag{3.7}$$

Thus, the direct problem can be solved by using erf.m, or more conveniently, by using **nf2.m** which takes care of the $\sqrt{2}$ factor automatically since $nf2(r) = \operatorname{erf}(r/\sqrt{2})$. Because nf2.m is a MATLAB function file, it can be called in several different but equivalent ways:

- `r= 0.75; p = nf2(r)`
- `nf2(0.75)`
- `p = nf2(0.75)`

EXAMPLE 3.1

Suppose it is known that the measurement error from using a yardstick is normally distributed with zero mean and a standard deviation of 5 mm. What is the probability that a particular measurement made with the stick is in error by ± 3 mm?

$$\text{Since } R = 3\,\text{mm}, r = R/\sigma = 3/5 = 0.6,$$

$$p = nf2(0.6) = 45.15\%$$

Notice that we first divide 3 by 5 to normalize the problem, so that the normalized half-interval length is $R/\sigma = 0.6$.

Figure 3.4 shows the PDF $h(\tau, \sigma)$ with $\sigma = 1$. The dotted area under the $h(\tau, 1)$ curve in the interval $[-0.6, 0.6]$ (or $R = 0.6$) is the probability $p = H(0.6) = 0.4515$.

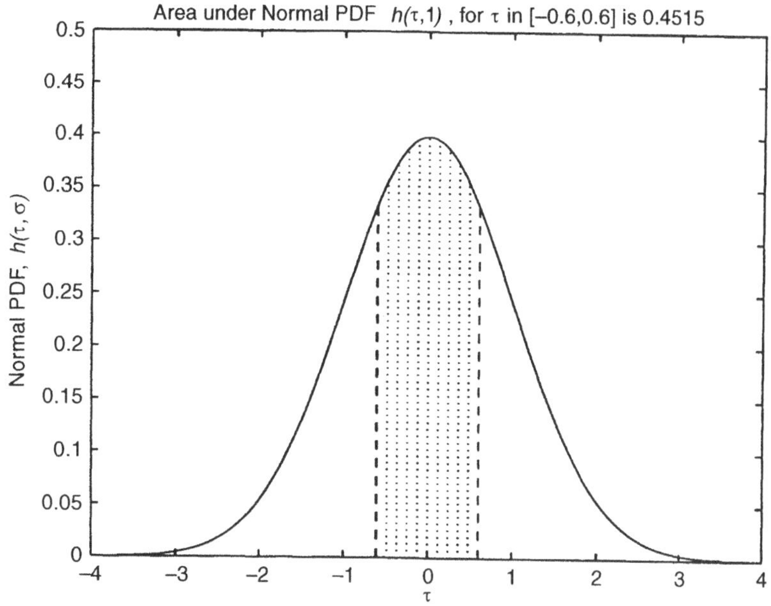

Figure 3.4 $p = 2\int_0^{0.6} h(\tau, \sigma)\, dt$ for the Normal PDF, $\sigma = 1$.

However, if $\sigma = 2$, then $h(\tau, 2)$ will be different from that in Figure 3.4 and we have $p = H(0.6/2) = \text{erf}(0.6/\sqrt{2}) = 0.2358$, according to Equation (3.7).

Figure 3.4 is generated with the MATLAB command *nf2a*(1). The program **nf2a.m** is similar to nf2.m, except that it produces a graph to explain the result in addition to generating a numerical answer.

Table 3.1 lists the R-to-p pairs, with three forms of R ($k \cdot \sigma$, $k \cdot$ MAE, $k \cdot$ LEP), for estimating the approximate solution of the direct problem. Here $p = H(r) = H(R/\sigma)$ from Equation (3.7).

3.4 INVERSE PROBLEMS

For the inverse problem, the goal is to find R or r from Equation (3.5) or Equation (3.6) such that

$$p = H(R/\sigma) \quad \text{or} \quad p = H(r), \text{ with } p \text{ given}$$

Taking the inverse on both sides of Equation (3.7), we obtain

$$\text{erfinv}(p) = r/\sqrt{2}$$

or

$$r = \sqrt{2}\, \text{erfinv}(p) \tag{3.8}$$

TABLE 3.1 One-Dimensional Direct Problem, from R to p

k	$p = H(R/\sigma)$ $R = k \cdot \sigma$	$p = H(R/\sigma)$ $R = k \cdot \text{MAE}$	$p = H(R/\sigma)$ $R = k \cdot \text{LEP}$
0.0	0.0000	0.0000	0.0000
0.1	0.0797	0.0636	0.0538
0.2	0.1585	0.1268	0.1073
0.3	0.2358	0.1892	0.1604
0.4	0.3108	0.2504	0.2127
0.5	0.3829	0.3101	0.2641
0.6	0.4515	0.3679	0.3143
0.7	0.5161	0.4235	0.3632
0.8	0.5763	0.4767	0.4105
0.9	0.6319	0.5273	0.4562
1.0	0.6827	0.5751	0.5000
1.1	0.7287	0.6199	0.5419
1.2	0.7699	0.6617	0.5817
1.3	0.8064	0.7004	0.6194
1.4	0.8385	0.7360	0.6550
1.5	0.8664	0.7686	0.6883
1.6	0.8904	0.7983	0.7195
1.7	0.9109	0.8250	0.7485
1.8	0.9281	0.8491	0.7753
1.9	0.9426	0.8705	0.8000
2.0	0.9545	0.8895	0.8227
2.1	0.9643	0.9062	0.8434
2.2	0.9722	0.9208	0.8622
2.3	0.9786	0.9335	0.8792
2.4	0.9836	0.9445	0.8945
2.5	0.9876	0.9539	0.9083
2.6	0.9907	0.9620	0.9205
2.7	0.9931	0.9688	0.9314
2.8	0.9949	0.9745	0.9411
2.9	0.9963	0.9793	0.9495
3.0	0.9973	0.9833	0.9570
3.1	0.9981	0.9866	0.9635
3.2	0.9986	0.9893	0.9691
3.3	0.9990	0.9915	0.9740
3.4	0.9993	0.9933	0.9782
3.5	0.9995	0.9948	0.9818
3.6	0.9997	0.9959	0.9848
3.7	0.9998	0.9968	0.9874
3.8	0.9999	0.9976	0.9896
3.9	0.9999	0.9981	0.9915
4.0	0.9999	0.9986	0.9930

The M-file **nf3.m** is written to perform the operations indicated in Equation (3.8). For any p in [0,1], nf3.m can be called in one of the following ways:

- `p=0.5; r=nf3(p)`
- `nf 3(0.5)`
- `r=nf3(0.5)`

EXAMPLE 3.2

Given a sample of 50 test errors, assume the population is normally distributed. What is the interval in which a test error will fall within 80% of the time? (A 50-sample set of error data is included in the file eg3p2.mat.)

Enter these commands in MATLAB

```
.load \mfile \eg 3p2.mat
mu = mean(z)
sig = std(z)
r = nf3(0.8)
```

It is found that $\mu_z = 2.0933$, $\sigma_z = 0.9592$, and $r = 1.2816$. Since $\mathbf{x} \sim N(0, 1)$ and z is related to x by $x = (z - \mu_z)/\sigma_z$, then $z = x\sigma_z + \mu_z$. Now that x is in the interval $[-1.2816, 1.2816]$, z is therefore in $[0.8641, 3.323]$. The program **nf7.m** carries out the conversion just mentioned and displays the result graphically in Figure 3.5. Note that the probability density function for $\mathbf{z}$ is $N(\mu_z, \sigma_z^2) = N(2.0933, 0.9592^2)$.

For direct problems, the M-file **r2p1d.m** takes as input a multiple (k) of SIGMA (σ), MAE, and LEP, and then produces the probability $p = H(k \cdot \sigma/\sigma) = H(k)$, $p = H(k \cdot \text{MAE}/\sigma)$, and $p = F(k \cdot \text{LEP}/\sigma)$, respectively. For example, entering r2p1d(2) in MATLAB results in

```
For R = 2 * SIGMA
  p = 0.9545
For R = 2 * MAE
  p = 0.8895
For R = 2 * LEP
  p = 0.8227
```

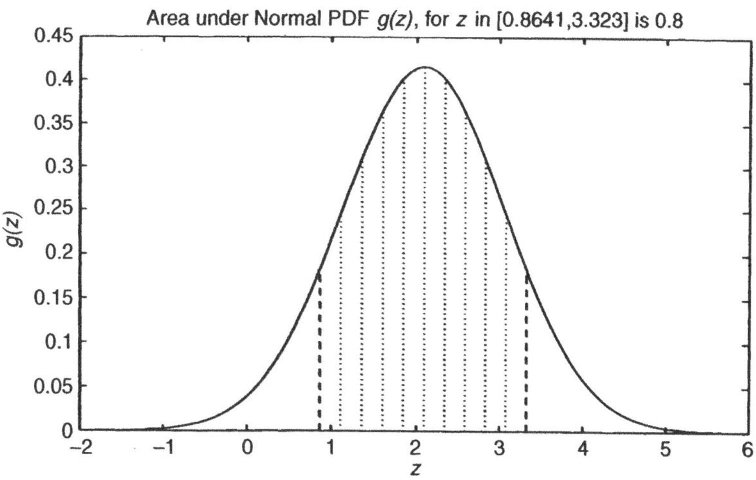

Figure 3.5 Solution for Example 3.2.

For inverse problems, the M-file **p2r1d.m** takes p as input and generates r expressed in terms of SIGMA (σ), MAE, and LEP. This may be considered as a conversion tool between these three types of one-dimensional errors. For example, entering p2r1d(0.5) in MATLAB results in

```
For p = 0.5
R = 0.6745 * SIGMA
R = 0.8453 * MAE
R = 1 * LEP
```

TABLE 3.2 One-Dimensional Inverse Problem, from p to R

p	R/σ	R/MAE	R/LEP
0.000	0.0000	0.0000	0.0000
0.025	0.0313	0.0393	0.0465
0.050	0.0627	0.0786	0.0930
0.075	0.0941	0.1180	0.1396
0.100	0.1257	0.1575	0.1863
0.125	0.1573	0.1972	0.2332
0.150	0.1891	0.2370	0.2804
0.175	0.2211	0.2771	0.3278
0.200	0.2533	0.3175	0.3756
0.225	0.2858	0.3582	0.4238
0.250	0.3186	0.3994	0.4724
0.275	0.3518	0.4409	0.5216
0.300	0.3853	0.4829	0.5713
0.325	0.4193	0.5255	0.6216
0.350	0.4538	0.5687	0.6727
0.375	0.4888	0.6126	0.7247
0.400	0.5244	0.6572	0.7775
0.425	0.5607	0.7027	0.8313
0.450	0.5978	0.7492	0.8862
0.475	0.6357	0.7967	0.9424
0.500	0.6745	0.8453	1.0000
0.525	0.7144	0.8953	1.0591
0.550	0.7554	0.9468	1.1200
0.575	0.7978	0.9999	1.1828
0.600	0.8416	1.0548	1.2478
0.625	0.8871	1.1119	1.3153
0.650	0.9346	1.1713	1.3856
0.675	0.9842	1.2336	1.4592
0.700	1.0364	1.2990	1.5366
0.725	1.0916	1.3681	1.6184
0.750	1.1503	1.4417	1.7055
0.775	1.2133	1.5207	1.7989
0.800	1.2816	1.6062	1.9000
0.825	1.3563	1.6999	2.0109
0.850	1.4395	1.8042	2.1343
0.875	1.5341	1.9227	2.2745
0.900	1.6449	2.0615	2.4387
0.925	1.7805	2.2315	2.6397
0.950	1.9600	2.4565	2.9058
0.975	2.2414	2.8092	3.3231
0.990	2.5758	3.2283	3.8189

Table 3.2 lists the p-to-R pairs for estimating the approximate solution of the inverse problem for readers who prefer table look-up to the use of MATLAB.

3.5 SUMMARY OF M-FILES

The M-files used or generated in this chapter are summarized in Table 3.3. Recall that the half-interval length R and the normalized half-interval length r are related by

$$r = R/\sigma$$

TABLE 3.3 M-files Used in Chapter 3

File Name	MATLAB Command
nf1.m	$y = nf1(x)$
nf2.m	$p = nf2(r)$
nf3.m	$r = nf3(p)$
nf6.m	$y = nf6(x, \mu, \sigma)$
nf7.m	$nf7(0.8)$
r2p1d.m	$r2p1d(k)$
p2r1d.m	$p2r1d(p)$

The standard normal distribution is implemented in the M-file **nf1.m**. The M-file **nf6.m** takes (x, μ, σ) as inputs and generates

$$y = \frac{1}{\sqrt{2\pi}\sigma} \exp\left[-\frac{(x-\mu)^2}{2\sigma^2}\right]$$

as output. The file nf6.m is invoked in the program nf7.m.

4

Two-Dimensional Error Analysis

Errors in the measurement of distance on a planar or near-planar region are treated in this chapter. This type of error is the result of a root sum square process of two one-dimensional errors in two different channels.

The joint probability distribution of the angular position for a random point in the plane is considered at the end of this chapter.

4.1 TWO-DIMENSIONAL NORMAL DISTRIBUTION

In two-dimensional error analysis, the most general **bivariate normal** joint probability density function of random variables (**x**, **y**) is

$$g(x, y) = \frac{1}{2\pi\sigma_x \sigma_y \sqrt{1-\rho^2}} \times \exp\left\{\frac{-1}{2(1-\rho^2)}\left[\frac{(x-\mu_x)^2}{\sigma_x^2} - \frac{2\rho(x-\mu_x)(y-\mu_y)}{\sigma_x \sigma_y} + \frac{(y-\mu_y)^2}{\sigma_y^2}\right]\right\}$$

where μ_x and σ_x are the mean and standard deviation of the random variable **x**; μ_y and σ_y are the mean and standard deviation of the random variable **y**; and ρ is the **correlation coefficient** between the two one-dimensional random variables **x** and **y**.

After removing the means μ_x and μ_y, the joint probability density function is simplified to

$$f(x, y) = \frac{1}{2\pi\sigma_x \sigma_y \sqrt{1-\rho^2}} \exp\left\{\frac{-1}{2(1-\rho^2)}\left[\frac{x^2}{\sigma_x^2} - \frac{2\rho xy}{\sigma_x \sigma_y} + \frac{y^2}{\sigma_y^2}\right]\right\} \tag{4.1}$$

Define the **ellipticity** u, the ratio of the smaller of the two standard deviations (σ_x, σ_y) to the larger one, as

$$u = \frac{\min(\sigma_x, \sigma_y)}{\max(\sigma_x, \sigma_y)}$$

Throughout this chapter we assume, without loss of generality, that $\sigma_x \geq \sigma_y$; hence, $u = \sigma_y/\sigma_x$.

When (1) $\rho = 0$ and $u = 1$ (or $\sigma_x = \sigma_y$), the distribution is called a **circular normal distribution**, because the contours of equal probability density are concentric circles. When (2) $\rho \neq 0$ and $u \neq 1$, or (3) $\rho = 0$ and $u \neq 1$, or (4) $\rho \neq 0$ and $u = 1$, the distribution function is called an **elliptical normal distribution**, because the contours of equal probability density are confocal ellipses. Thus, ellipticity alone does not determine the shape of the contour, since we must also look at the value of the correlation coefficient. Figure 4.1 illustrates the four different situations for $1/(1-\rho^2)[x^2/\sigma_x^2 - 2\rho xy/(\sigma_x\sigma_y) + y^2/\sigma_y^2] = 1$.

Figure 4.2 shows the wire plot of an elliptical normal distribution, $z = f(x, y)$, with $\sigma_x = 2$, $\sigma_y = \sqrt{3}$, and $\rho = 0.4$. The corresponding contours of equal probability density are plotted in Figure 4.3.

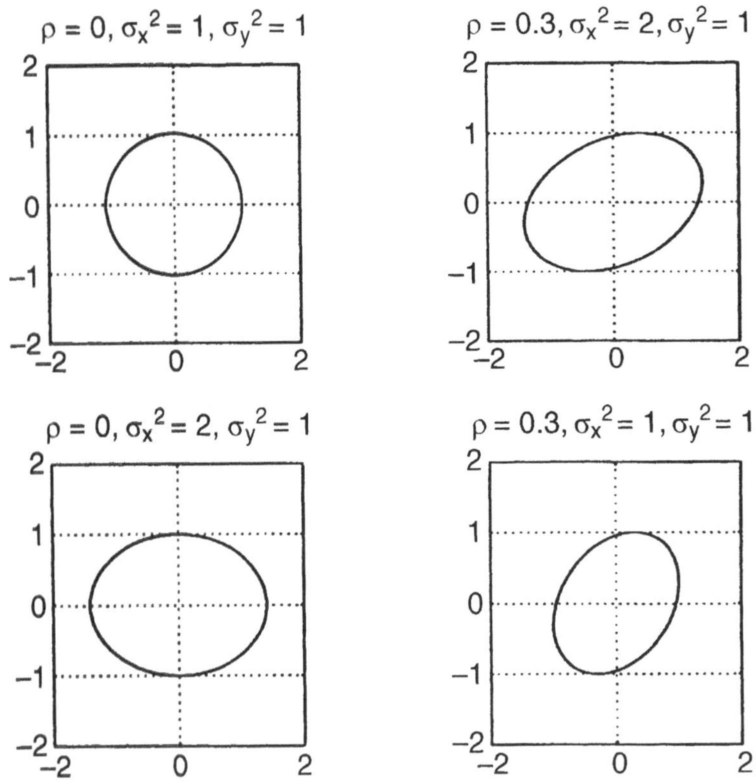

Figure 4.1 Constant PDF Contours of Circular and Elliptical Normal Distributions.

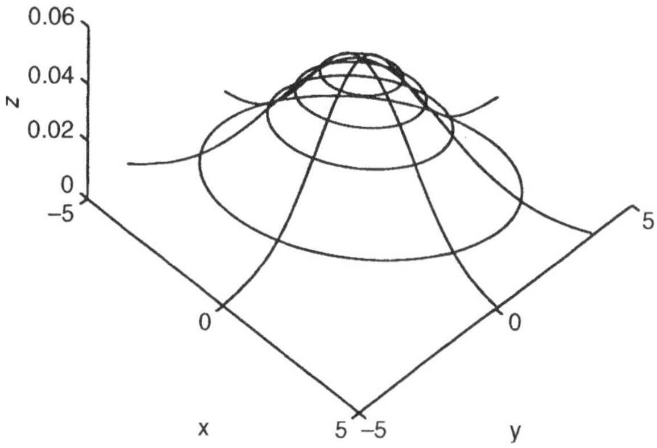

Figure 4.2 Two-Dimensional Elliptical Normal Distribution.

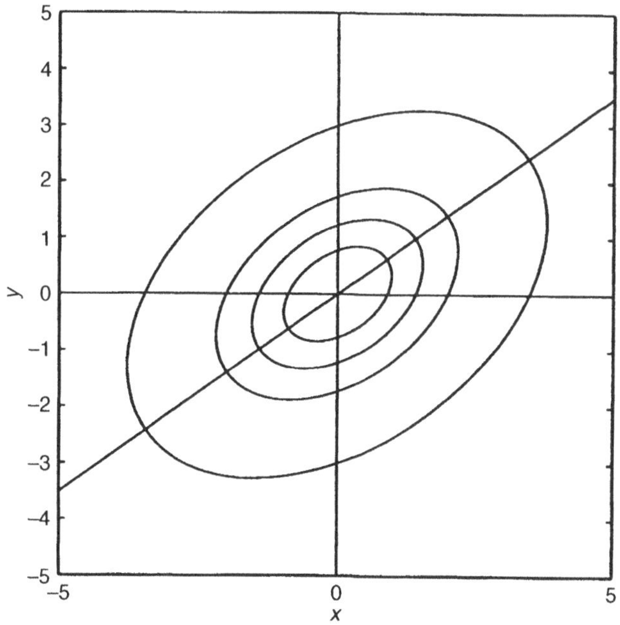

Figure 4.3 Contours of Equal Probability Density.

4.2 DIRECT AND INVERSE PROBLEMS

The goal of the direct problem in two-dimensional error analysis is to find the probability by evaluating the double integral of the function $f(x, y)$ in Equation (4.1), over a given planar region A, as

$$p = \iint_A \frac{1}{2\pi\sigma_x\sigma_y\sqrt{1-\rho^2}} \exp\left\{\frac{-1}{2(1-\rho^2)}\left[\frac{x^2}{\sigma_x^2} - \frac{2\rho xy}{\sigma_x\sigma_y} + \frac{y^2}{\sigma_y^2}\right]\right\} dx\,dy \qquad (4.2)$$

where the integration region A can be one of three types:

1. Rectangular region $A = A_r\colon |x| \leq a,\ |y| \leq b$
2. Circular region $A = A_c\colon x^2 + y^2 \leq R^2$
3. Elliptical region $A = A_e\colon (x/a)^2 + (y/b)^2 \leq 1$

For inverse problems we start with a specified probability p_0 and then proceed to find the dimensions of the integration region—that is, the width and length of a rectangle, the radius of a circle, or the semi-major and semi-minor axes of an ellipse. The solution is found via a repetitive trial-and-comparison procedure as:

(1) Assume an initial dimension(s) for A; then choose a reasonably small tolerance limit δ.
(2) Carry out the double integration in Equation (4.2) to obtain p.
(3) Compare p with p_0.
(4) If $|p - p_0| < \delta$, accurate dimensions of A have been found, stop.
(5) If $p - p_0 > \delta$, decrease the dimensions of A, then go back to (2).
(6) If $p_0 - p > \delta$, increase the dimensions of A, then go back to (2).

Therefore, both the direct problem and the inverse problem require an efficient way to compute the double integral of Equation (4.2).

Depending on the size of the ellipticity u, the shape of the integration region A, and the value of the correlation coefficient ρ, six cases are to be investigated:

- Case 1, $A = A_r,\ \rho = 0,\ u \in [0, 1]$
- Case 2, $A = A_c,\ \rho = 0,\ u = 1$
- Case 3, $A = A_c,\ \rho = 0,\ u \in [0, 1]$
- Case 4, $A = A_c,\ \rho \in [-1, 1],\ u \in [0, 1]$
- Case 5, $A = A_e,\ \rho = 0,\ u \in [0, 1]$
- Case 6, $A = A_e,\ \rho \in [-1, 1],\ u \in [0, 1]$

In order to easily refer to these cases without lengthy description, we rely on an abbreviation system using six characters (XXYYZP). The meaning of the six-character string is as follows:

XX: (ED, CD)	(Elliptical Distribution, Circular Distribution)
YY: (RA, CA, EA)	(Rectangular, Circular, Elliptical Area)
Z: (U, C)	(Uncorrelated, Correlated)
P: (D, I)	(Direct problem, Inverse problem)

Thus, the direct and inverse problems to be considered are:

Case 1 (EDRAUD), (EDRAUI)
Case 2 (CDCAUD), (CDCAUI)
Case 3 (EDCAUD), (EDCAUI)
Case 4 (EDCACD), (EDCACI)
Case 5 (EDEAUD), (EDEAUI)
Case 6 (EDEACD), (EDEACI)

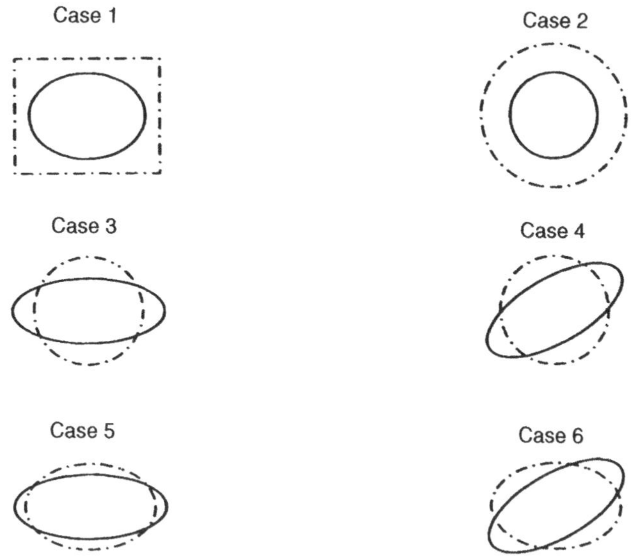

Figure 4.4 Equal PDF Contour (solid line) and Boundary of Integration Region (dashed line).

These cases are studied in Sections 4.3 through 4.9. Figure 4.4 shows a representative geometry (equal PDF contour and integration region) for each of the six cases. When naming M-files developed for these six cases, we will also use the same abbreviation convention.

4.3 CASE 1, $A = A_r$, $\rho = 0$, $u \in [0, 1]$

With the assumptions that the region of integration is rectangular and the two channels are uncorrelated, the double integral in Equation (4.2) can be written as

$$p = \int_{-b}^{b} \int_{-a}^{a} \frac{1}{2\pi\sigma_x \sigma_y} \exp\left\{\frac{-1}{2}\left[\frac{x^2}{\sigma_x^2} + \frac{y^2}{\sigma_y^2}\right]\right\} dx\, dy \tag{4.3}$$

which is then simplified to

$$p = \operatorname{erf}\left[\frac{a}{\sqrt{2}\,\sigma_x}\right] \cdot \operatorname{erf}\left[\frac{b}{\sqrt{2}\,\sigma_y}\right] \tag{4.4}$$

Even though four parameters (a, b, σ_x, σ_y) appear in the preceding equation, only two normalized variables ($a_0 = a/\sigma_x$, $b_0 = b/\sigma_y$) are needed to determine the probability p. The direct problem can be solved either from look-up in Table 4.1, containing the values of $\operatorname{erf}(R_0/\sqrt{2})$, or the M-file **edraud.m** can be used. The MATLAB command to use this file is $p = edraud(a, b, \sigma_x, \sigma_y)$. Upon using Table 4.1, we should first split R_0 into an integer part i and a decimal part d, that is, $R_0 = i + d$. The value of $\operatorname{erf}(R_0/\sqrt{2})$ is then obtained at the intersection of column i and row d. Thus, if $R_0 = 1.2$, we have $i = 1$, $d = 0.2$, and $\operatorname{erf}(1.2/\sqrt{2}) = 0.7699$. We can also use the M-file **nf2.m** via the command $nf2(1.2)$ to arrive at the same answer, since $\operatorname{erf}(r/\sqrt{2}) = nf2(r)$.

TABLE 4.1 Values of $\mathrm{erf}(R_0/\sqrt{2})$

$R_0 = i + d$	i			
d	0	1	2	3
0.0	0.0000	0.6827	0.9545	0.9973
0.1	0.0797	0.7287	0.9643	0.9981
0.2	0.1585	0.7699	0.9722	0.9986
0.3	0.2358	0.8064	0.9786	0.9990
0.4	0.3108	0.8385	0.9836	0.9993
0.5	0.3829	0.8664	0.9876	0.9995
0.6	0.4515	0.8904	0.9907	0.9997
0.7	0.5161	0.9109	0.9931	0.9998
0.8	0.5763	0.9281	0.9949	0.9999
0.9	0.6319	0.9426	0.9963	0.9999

An example follows.

EXAMPLE 4.1

If a position error has an elliptical normal PDF with $\sigma_x = 0.4$ nmi and $\sigma_y = 1$ nmi, and errors in the x, y channels are uncorrelated, what is the probability that a measured position error will fall within a rectangle bounded by $|x| \leq a = 0.6$ nmi and $|y| \leq b = 1.2$ nmi?

The probability can be found from Table 4.1, with $a_0 = (0.6/0.4) = 1.5$ and $b_0 = (1.2/1) = 1.2$. Hence, $p = \mathrm{erf}(a_0/\sqrt{2})\mathrm{erf}(b_0/\sqrt{2}) = nf2(a_0)nf2(b_0) = 0.8664 \times 0.7699 = 0.6670$ or enter the MATLAB command:

```
p = edraud(0.6, 1.2, 0.4, 1)
```

which gives the same result.

Given σ_x, σ_y, and p, there are many (a, b) pairs that satisfy Equation (4.4). The inverse problem can result in multiple solutions of (a, b) if one does not specify the ratio $v = a/b$. Notice that, unlike the ratio u which is bounded by 0 and 1, the ratio v can have any value in $[0, \infty)$. Solutions for inverse problems are obtainable from **edraui.m**. The MATLAB command is $[a, b] = edraui(p, v, \sigma_x, \sigma_y)$. Let us look at another example.

EXAMPLE 4.2

The joint PDF of a certain position error is elliptical normal with $\sigma_x = 1.5$ nmi and $\sigma_y = 0.6$ nmi, and errors in the x, y channels are uncorrelated. Determine length a along the x-axis and width b along the y-axis, if the specified ratio is $v = (a/b) = 2$, and also if the probability is 90% for a sample position error to fall inside this $2a \times 2b$ rectangle.

After entering the MATLAB command

```
[a,b] = edraui(0.9,2.0,1.5,0.6),
```

we obtain $a = 2.6650$, $b = 1.3325$. Note that the requirement $v = 2 = (a/b) = (2.6650/1.3325)$ is met.

If $v = (a/b) = 1/2$, we obtain, from

```
[a,b] = edraui(0.9,0.5,1.5,0.6)
```

the alternate results $a = 2.4673$, $b = 4.9346$. This demonstrates the necessity of clearly specifying $v = a/b$ for the inverse problem.

4.3.1 Limiting Situation: $u \to 0$

As the ellipticity u approaches zero, the two-dimensional joint distribution becomes that of a one-dimensional PDF. We have in place of Equation (4.4):

$$p = \int_{-a}^{a} \frac{1}{\sqrt{2\pi}\,\sigma_x} \exp\left[\frac{-x^2}{2\sigma_x^2}\right] dx = \text{erf}\left[\frac{a}{\sqrt{2}\,\sigma_x}\right] \tag{4.5}$$

4.4 CASE 2, $A = A_c$, $\rho = 0$, $u = 1$

When the random variables $\mathbf{x}$, $\mathbf{y}$ are uncorrelated ($\rho = 0$) and have equal standard deviation ($\sigma_x = \sigma_y = \sigma$), and the integration region is circular with radius R, Equation (4.2) becomes

$$p = \iint_{A_c} \frac{1}{2\pi\sigma^2} \exp\left[-\frac{1}{2\sigma^2}(x^2 + y^2)\right] dx\,dy \tag{4.6}$$

with A_c: $x^2 + y^2 \leq R^2$. This double integral can be transformed to an equivalent polar form

$$p = \int_0^{2\pi}\int_0^R \frac{1}{2\pi\sigma^2} \exp\left[\frac{-r^2}{2\sigma^2}\right] r\,dr\,d\theta = \int_0^R h(r, \sigma)\,dr = 1 - \exp\left[\frac{-R^2}{2\sigma^2}\right] \tag{4.7}$$

Note that

$$h(r, \sigma) = \frac{r}{\sigma^2} \exp\left[\frac{-r^2}{2\sigma^2}\right], \quad \text{for } r \geq 0 \tag{4.8}$$

which is the **Rayleigh distribution function**. The independent variable of this function corresponds, for example, to the random planar radial error, $\mathbf{r} = \sqrt{\mathbf{x}^2 + \mathbf{y}^2}$.

Following the notation convention of Chapter 3 (page 25), where we used the notation $h(r, \sigma)$ as the PDF of $\mathbf{r}$, and $H(R/\sigma) = 2\int_0^r h(\tau, \sigma)\,d\tau$ as the CDF, we now have $h(r, \sigma)$ and $H(R/\sigma) = \int_0^R h(t, \sigma)\,dt$ for PDF and CDF, respectively. Thus, with Equation (4.7) and the substitution of the normalized radius $r = R/\sigma$, we have

$$p = H(R/\sigma) = 1 - \exp(-R^2/2\sigma^2) = H(r) = 1 - \exp(-r^2/2) \tag{4.9}$$

Geometrically, this means that the volume under $f(x, y)$ over the region $x^2 + y^2 \leq R^2$ is equivalent to the area under the Rayleigh PDF $h(r, \sigma)$ over the interval $[0,\ R]$. Therefore in this special case, and this case alone, two-dimensional error analysis can be performed in terms of one random variable $\mathbf{r}$, the planar (two-dimensional) radial error.

Equation (4.9) provides the closed-form solution for the direct problem in this special case, and the MATLAB M-file for it is **cdcaud.m**. The command form in MATLAB is $p = cdcaud(R, \sigma)$. An example follows.

EXAMPLE 4.3

If a position error has a circular normal distribution with $\sigma_x = \sigma_y = 0.8$ nmi, and errors in the x, y channels are uncorrelated, what is the probability that a sample position error will fall within a circle of radius $R = 1.2$ nmi?

```
p = cdcaud(1.2,0.8) = 0.6753
```

There is a 67.53% chance.

The inverse problem can also be solved from Equation (4.9) analytically in this particular case. When p is given,

$$R = \sigma\sqrt{-2\ \ln(1-p)} \tag{4.10}$$

The MATLAB M-file **cdcaui.m** is implemented for solving this inverse problem. Its command form is $R = cdcaui(p, \sigma)$. An example is given next.

EXAMPLE 4.4

If position error has a circular normal distribution with $\sigma_x = \sigma_y = 1.8$ nmi, and the errors in the x, y channels are uncorrelated, find the radius of a circle such that 50% of the time a measured position error will fall inside this circle?

```
R = cdcaui(0.5, 1.8) = 2.1193
```

Thus, the radius of this 50%-probability circle is 2.1193 nmi. The radius for a $p = 50\%$ circle, $R_{50\%}$, is also called the CEP (see next section).

The M-files **cf1.m, cf2.m**, and **cf3.m** carry out the mathematics in Equations (4.8) through (4.10), respectively; cf1.m is for the Rayleigh PDF, cf2.m is for direct problems, while cf3.m is for inverse problems. They all assume $\sigma = 1$. Hence, cf2.m and cf3.m are the normalized counterparts of cdcaud.m and cdcaui.m, respectively. This means that the two commands $p = cdcaud(R, \sigma)$ and $R = cdcaui(p, \sigma)$ can be obtained from $p = cf2(r)$ and $r = cf3(p)$, with $r = R/\sigma$ being the **normalized radius**.

4.5 TWO-DIMENSIONAL ERROR MEASURES

Because of its simplicity, the Rayleigh distribution in Equation (4.8) will be used to illustrate several two-dimensional error indicators. These include mean radial error (MRE), root mean square radial error (RMSR), distance root mean square error (DRMS), standard radial error (SRE), and circular error probable (CEP, see **[8]**).

4.5.1 Mean Radial Error (MRE)

Mean radial error (**MRE**, μ_r) is defined as

$$\text{MRE} = \mu_r = E[\mathbf{r}] = \int_0^{\infty} rh(r,\sigma)\,dr = \sqrt{\frac{\pi}{2}}\sigma \approx 1.2533\sigma \tag{4.11}$$

4.5.2 Root Mean Square Radial Error (RMSR)

The mean square radial (**MSR**) error is defined as

$$\text{MSR} = E[\mathbf{r}^2] = \int_0^\infty r^2 h(r, \sigma)\, dr = 2\sigma^2$$

The root mean square radial error (**RMSR**), also called **radial error (RE)** in **[8]**, or **radial position error (RPE)**, is obtained by taking the square root of the MSR,

$$\text{RMSR} = \text{RE} = \text{RPE} = \sqrt{E[\mathbf{r}^2]} = \sqrt{2}\,\sigma \approx 1.414\sigma \tag{4.12}$$

To avoid confusion with the random variable introduced in Section 4.4, the author will not use the term *radial error* for the type of error shown above.

4.5.3 Distance Root Mean Square Error (DRMS)

DRMS is defined as the square root of the sum of the **x** and **y** variances as

$$\text{DRMS} = \sqrt{\sigma_x^2 + \sigma_y^2} = \sqrt{1 + u^2}\,\sigma_x \tag{4.13}$$

When $u = 1$ (or $\sigma_x = \sigma_y = \sigma$),

$$\text{DRMS} = \sqrt{2}\,\sigma = \text{RMSR}$$

but, in general, DRMS $\neq$ RMSR.

The GPS horizontal dilution of precision (HDOP) is defined in terms of this two-dimensional DRMS (see Appendix D).

4.5.4 Standard Radial Error (SRE)

The variance of the radial error **r** is

$$\sigma_r^2 = E[(\mathbf{r} - \mu_r)^2] = E[\mathbf{r}^2] - (\mu_r)^2 = \left(2 - \frac{\pi}{2}\right)\sigma^2 \tag{4.14}$$

The standard radial error is the square root of this variance; thus, SRE $= \sigma_r \approx 0.6551\sigma$.

These three error indicators (SRE, RMSR, MRE) are related as

$$\sigma_r^2 = (\text{SRE})^2 = (\text{RMSR})^2 - (\text{MRE})^2 \tag{4.15}$$

Notice that RMSR (or DRMS) is not equal to σ or σ_r.

4.5.5 Circular Error Probable (CEP)

The circular error probable (**CEP**) is defined to be the particular radius of a circular region A_c over which the double integral in Equation (4.6) produces a probability of 50%. Under these conditions, we see from Equation (4.10) that

$$\text{CEP} = R_{50\%} = \sigma\sqrt{-2\ln(1 - 0.5)} = \sigma\sqrt{2\ln(2)} = 1.1774\sigma$$

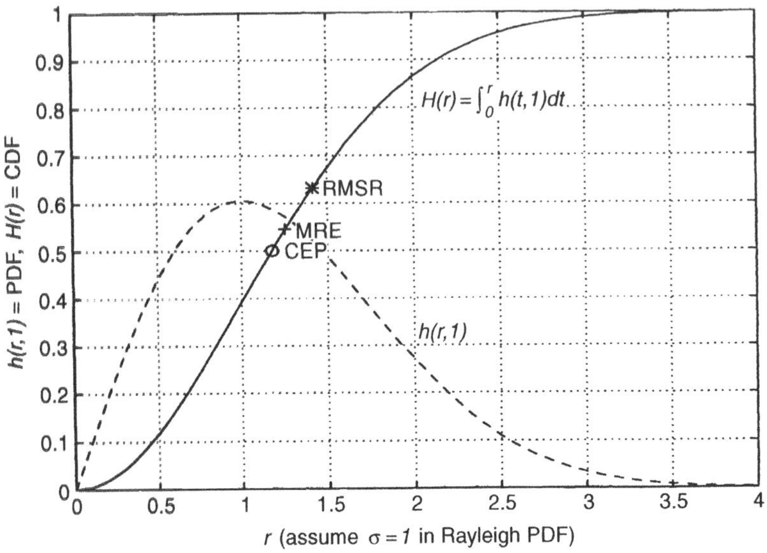

Figure 4.5 PDF and CDF for the Rayleigh Distribution.

We can also use one of the MATLAB commands $R = cdcaui(0.5, 1)$ or $r = cf3(0.5)$, to obtain 1.1774. Figure 4.5 shows the PDF $h(r, \sigma)$ with $\sigma = 1$, and the CDF $H(r)$ of a Rayleigh random variable.

The RMSR, MRE, and CEP are marked with "*", "+", and "o", respectively. Other points of interest are the 2σ and 3σ probabilities, which are 86.47% and 98.89%, respectively. Upon substituting $\sigma = \text{CEP}/\sqrt{2\ln(2)}$ into Equation (4.9), we have

$$p = 1 - 2^{-(R/\text{CEP})^2}$$

which is easier to use when CEP is given in place of σ.

Since the variable r in $H(r)$ is to be interpreted as a normalized quantity ($r = R/\sigma$), the solid CDF curve in Figure 4.5 is valid for all positive values of σ. The variable r in $h(r, \sigma)$ should not be treated as a relative quantity. Thus, the dashed PDF curve in Figure 4.5 is valid only for $\sigma = 1$. In other words, the variable used for the horizontal axis has two interpretations, one for the CDF and another for the PDF.

Figure 4.6 shows the PDF $h(r, \sigma)$ of Equation (4.8) with $\sigma = 1$. The shaded area under the $h(r, 1)$ curve in the [0,1] interval (or $R = 1$) is the probability $p = H(1/1) = 1 - \exp(-1/2) = 0.3935$, according to Equation (4.9). However, if $\sigma = 2$, then $h(r, 2)$ will be different than that in Figure 4.6 and we have $p = H(1/2) = 1 - \exp(-1/8) = 0.1175$.

Some military agencies use the term **dispersion** (δ) to mean the *diameter* (not radius) of a 75% probability circle where

$$\delta = \text{dispersion} = 2 * R_{75\%} = 2 * 1.6651\sigma = 3.3302\sigma$$

The terms 1-SIGMA (1σ) error, 2-SIGMA error, and so on, in two-dimensional error analysis should be used only for the special circular normal distribution case. In all other cases, it is not even defined. For example, a statement such as "the position accuracy of a system is 60 m (2σ)" could leave the reader wondering

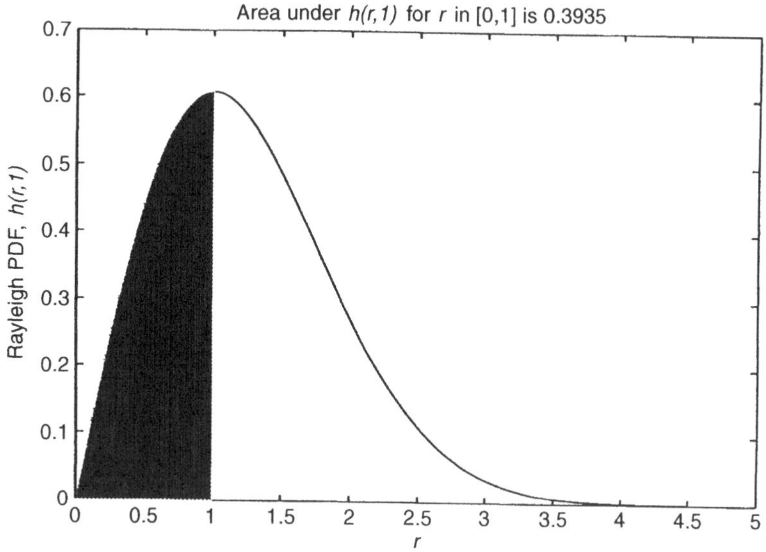

Figure 4.6 $p = \int_0^1 h(r, \sigma)\, dr$ for the Rayleigh PDF, $\sigma = 1$.

whether the distribution is circular, in which case (from the Rayleigh distribution) the numbers describe the 86.47% probability circle. It could also be interpreted as one-dimensional sigmas along each axis, in which case (from the standard normal distribution) the 95.45% probability circle is implied (assuming the distribution to be circular normal, which actually may not be so). Extra care is needed to avoid confusion.

Table 4.2 contains the R-to-p pairs for the direct problem in this special case. R is expressed in three forms: $R = k \cdot \sigma$, $R = k \cdot \text{MRE}$, and $R = k \cdot \text{CEP}$. Table 4.3 shows the p-to-R pairs for the inverse problem, where, for a given p, R is expressed as multiples of σ, MRE, and CEP; that is, $k = R/\sigma$, $k = R/\text{MRE}$, and $k = R/\text{CEP}$, respectively.

The probabilities for $R = 1$ DRMS and 2 DRMS are, respectively, 63.21% and 98.17%, see **[14]**.

Conversion between these error indicators is possible via the two M-files, **r2p2d.m** and **p2r2d.m**. The former (r2p2d.m) takes as input a multiple (k) of SIGMA, MRE, DRMS, and CEP, and then produces the corresponding probability p for $R = k \cdot \text{SIGMA}$, $R = k \cdot \text{MRE}$, $R = k \cdot \text{DRMS}$, and $R = k \cdot \text{CEP}$, respectively. For example, entering r2p2d(2) in MATLAB results in

```
For r = 2*SIGMA
  p = 0.8647
For R = 2*MRE
  p = 0.9568
For R = 2*DRMS(RMSR)
  p = 0.9817
For R = 2*CEP
  p = 0.9375
```

TABLE 4.2 Two-Dimensional Direct Problem, from R to p

k	$p = H(R/\sigma)$ $R = k \cdot \sigma$	$p = H(R/\sigma)$ $R = k \cdot \text{MRE}$	$p = H(R/\sigma)$ $R = k \cdot \text{CEP}$
0.0	0.0000	0.0000	0.0000
0.1	0.0050	0.0078	0.0069
0.2	0.0198	0.0309	0.0273
0.3	0.0440	0.0682	0.0605
0.4	0.0769	0.1181	0.1050
0.5	0.1175	0.1783	0.1591
0.6	0.1647	0.2463	0.2208
0.7	0.2173	0.3194	0.2880
0.8	0.2739	0.3951	0.3583
0.9	0.3330	0.4707	0.4296
1.0	0.3935	0.5441	0.5000
1.1	0.4539	0.6134	0.5677
1.2	0.5132	0.6773	0.6314
1.3	0.5704	0.7348	0.6901
1.4	0.6247	0.7855	0.7430
1.5	0.6753	0.8292	0.7898
1.6	0.7220	0.8661	0.8304
1.7	0.7643	0.8967	0.8651
1.8	0.8021	0.9215	0.8942
1.9	0.8355	0.9413	0.9181
2.0	0.8647	0.9568	0.9375
2.1	0.8897	0.9687	0.9530
2.2	0.9111	0.9777	0.9651
2.3	0.9290	0.9843	0.9744
2.4	0.9439	0.9892	0.9815
2.5	0.9561	0.9926	0.9869
2.6	0.9660	0.9951	0.9908
2.7	0.9739	0.9967	0.9936
2.8	0.9802	0.9979	0.9956
2.9	0.9851	0.9986	0.9971
3.0	0.9889	0.9991	0.9980
3.1	0.9918	0.9995	0.9987
3.2	0.9940	0.9997	0.9992
3.3	0.9957	0.9998	0.9995
3.4	0.9969	0.9999	0.9997
3.5	0.9978	0.9999	0.9998
3.6	0.9985	1.0000	0.9999
3.7	0.9989	1.0000	0.9999
3.8	0.9993	1.0000	1.0000
3.9	0.9995	1.0000	1.0000
4.0	0.9997	1.0000	1.0000

TABLE 4.3 Two-Dimensional Inverse Problem, from p to R

p	$k = R/\sigma$	$k = R/\text{MRE}$	$k = R/\text{CEP}$
0.000	0.0000	0.0000	0.0000
0.025	0.2250	0.1795	0.1911
0.050	0.3203	0.2556	0.2720
0.075	0.3949	0.3151	0.3354
0.100	0.4590	0.3663	0.3899
0.125	0.5168	0.4123	0.4389
0.150	0.5701	0.4549	0.4842
0.175	0.6203	0.4949	0.5268
0.200	0.6680	0.5330	0.5674
0.225	0.7140	0.5697	0.6064
0.250	0.7585	0.6052	0.6442
0.275	0.8020	0.6399	0.6811
0.300	0.8446	0.6739	0.7173
0.325	0.8866	0.7074	0.7530
0.350	0.9282	0.7406	0.7883
0.375	0.9695	0.7736	0.8235
0.400	1.0108	0.8065	0.8585
0.425	1.0520	0.8394	0.8935
0.450	1.0935	0.8725	0.9287
0.475	1.1352	0.9058	0.9642
0.500	1.1774	0.9394	1.0000
0.525	1.2202	0.9736	1.0363
0.550	1.2637	1.0083	1.0733
0.575	1.3082	1.0438	1.1111
0.600	1.3537	1.0801	1.1498
0.625	1.4006	1.1175	1.1896
0.650	1.4490	1.1561	1.2307
0.675	1.4993	1.1963	1.2734
0.700	1.5518	1.2381	1.3179
0.725	1.6069	1.2821	1.3647
0.750	1.6651	1.3286	1.4142
0.775	1.7272	1.3781	1.4670
0.800	1.7941	1.4315	1.5238
0.825	1.8671	1.4897	1.5857
0.850	1.9479	1.5542	1.6544
0.875	2.0393	1.6272	1.7321
0.900	2.1460	1.7122	1.8226
0.925	2.2761	1.8160	1.9331
0.950	2.4477	1.9530	2.0789
0.975	2.7162	2.1672	2.3069
0.990	3.0349	2.4215	2.5776

The latter (p2r2d.m) takes p as input and generates R expressed in terms of SIGMA, MRE, DRMS, and CEP. For example, entering p2r2d(0.5) in MATLAB results in

```
For p = 0.5

  r = 1.177*SIGMA

  r = 0.9394*MRE

  r = 0.8326*DRMS (RMSR)

  r = 1*CEP
```

This is a form of the inverse problem.

4.5.6 CEP Rate and RPE Rate

The time rates of change of CEP and RPE are called **CEPR** and **RPER**, respectively.

Under the assumption that $\sigma_x = \sigma_y = \sigma$, we have CEP $= 1.1774\sigma$ and RPE $= \sqrt{2}\sigma$ from Subsections 4.5.2 and 4.5.5. This means

$$\text{CEP} = 1.1774\frac{\text{RPE}}{\sqrt{2}} = 0.8325 \text{ RPE}$$

and thus the relationship between CEPR and RPER is

$$\text{CEPR} = 0.8325 \text{ RPER}$$

when we divide the equation, CEP $= 0.8325$ RPE, by unit time.

EXAMPLE 4.5

If the circular error on 95% of system flights is increasing at a rate of 2 nmi/hr, what are the rates of CEP and RPE?

Since $R_{95\%} = 2.4477\sigma = 2$ nmi, we can find $\sigma = 2/2.4477 = 0.817$ nmi. Thus, CEP $= R_{50\%} = 1.1774\sigma = 0.9620$ nmi, and RPE $= \sqrt{2}\sigma = 1.1555$ nmi. The corresponding rates are

$$\text{CEPR} = 0.9629 \text{ nmi/hr}$$

$$\text{RPER} = 1.1555 \text{ nmi/hr}$$

4.6 CASE 3, $A = A_c$, $\rho = 0$, $u \in [0, 1]$

In this case, the probability in Equation (4.2) becomes

$$p = \iint_{A_c} \frac{1}{2\pi\sigma_x \sigma_y} \exp\left\{\frac{-1}{2}\left[\frac{x^2}{\sigma_x^2} + \frac{y^2}{\sigma_y^2}\right]\right\} dx\, dy \tag{4.16}$$

where the area of interest A_c is a circle of radius R,

$$x^2 + y^2 \leq R^2$$

After we substitute $u = \sigma_y/\sigma_x$, $\lambda = r/\sigma_x = \sqrt{x^2 + y^2}/\sigma_x$, and the rectangular-to-polar transformation $x = r\cos\theta$, $y = r\sin\theta$, into the above equation, it becomes

$$p = \frac{1}{2\pi u} \int_0^{2\pi} \int_0^{R/\sigma_x} \exp\left[\frac{-\lambda^2}{2} S(\theta, u)\right] \lambda\, d\lambda\, d\theta \tag{4.17}$$

where $S(\theta, u) = \cos^2\theta + \sin^2\theta/u^2$. Integrating with respect to λ, we obtain

$$p = \frac{1}{2\pi u} \int_0^{2\pi} \frac{\{1 - \exp[-(R/\sigma_x)^2 S(\theta, u)/2]\}}{S(\theta, u)} d\theta \tag{4.18}$$

One can also perform integration with respect to θ in Equation (4.17), and apply the relationship

$$I_o(t) = \frac{1}{\pi} \int_0^{\pi} \exp[\pm t\cos\theta]\, d\theta = \frac{1}{2\pi} \int_0^{2\pi} \exp[\pm t\cos(2\theta)]\, d\theta$$

to obtain

$$p = \frac{1}{u} \int_0^{R/\sigma_x} \lambda \exp\left[\frac{-\lambda^2(1 + u^2)}{4u^2}\right] I_o\left[\frac{\lambda^2(1 - u^2)}{4u^2}\right] d\lambda \tag{4.19}$$

in terms of $I_o(t)$, which is the zero order modified Bessel function of the first kind; see **[12]**. Equation (4.18) is easier to use than Equation (4.19) for two reasons. First, it is a definite integral; second, the integrand function does not involve the modified Bessel function I_0. The MATLAB M-file **gf2.m** carries out the integration in Equation (4.18) for $u \in (0, 1)$ via Simpson's rule (which is implemented in **simprule.m**).

Thus, note that the probability p depends not only on the normalized radius R/σ_x, but also on the ellipticity $u = \sigma_y/\sigma_x$, which is the ratio of the smaller standard deviation to the larger standard deviation and is also a derivable or known parameter.

Numerical integration and root searching techniques are needed to solve both the direct and inverse problems. They are implemented in the M-files **edcaud.m** and **edcaui.m**.

Notice that when $u = 1$, case 3 specializes to case 2. As will be seen in the next few sections, cases 4–6 are transformable to case 3 by rotation or scaling. Therefore, case 3 is the heart of this chapter.

4.6.1 Limiting Situation: $u \to 0$

As u approaches 0, Equation (4.16) becomes

$$p = \int_{-R}^{R} \frac{1}{\sqrt{2\pi}\,\sigma_x} \exp\left[\frac{-x^2}{2\sigma_x^2}\right] dx = \text{erf}\left[\frac{R}{\sqrt{2}\,\sigma_x}\right]$$

The M-file **nf2.m** can be used to evaluate p. The MATLAB command is $p = nf2(r)$, where $r = R/\sigma_x$ is the normalized radius.

Figure 4.7 shows how the probability p varies when u is increased from 0 to 1, for the uncorrelated case corresponding to $\rho = 0$. The curve on the left ($R =$ 1 DRMS) indicates that p decreases as u increases; the one on the right ($R =$ 2 DRMS) displays a different behavior: p increases as u increases. Recall that as u varies, 1 DRMS $= \sqrt{1+u^2}\,\sigma_x$ does not remain constant. Later we will see a similar pattern for curves in the correlated, $\rho \neq 0$, case (Figure 4.11).

The normalized radius R/σ_x, as a function of ellipticity u, is plotted in Figure 4.8 for various probability values.

The curves in Figure 4.9 show the relationship between probability and radius for $u = 0$ to $u = 1$ in steps of 0.2. Note that the curve corresponding to $u = 0$ is identical to the one-dimensional normal CDF and the curve corresponding to $u = 1$ is identical to the Rayleigh CDF. Figure 4.10 repeats the bottom curve ($p = 0.5$) of Figure 4.8, along with the two approximations suggested by Pitman (see [10]):

$$\text{CEP}/\sigma_x = 0.589(1 + u)$$
$$\text{CEP}/\sigma_x = 0.562 + 0.615u$$

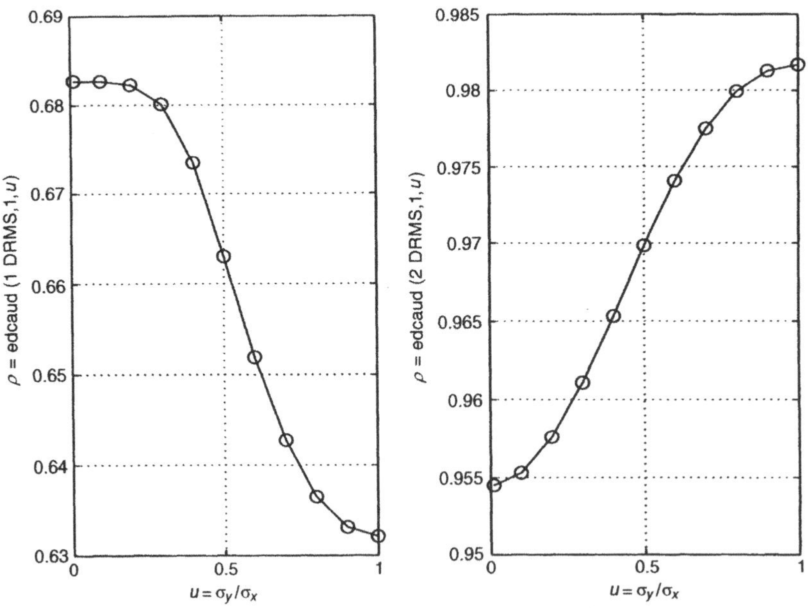

Figure 4.7 Probability Versus Ellipticity for $\rho = 0$ and $R =$ 1 DRMS and $R =$ 2 DRMS.

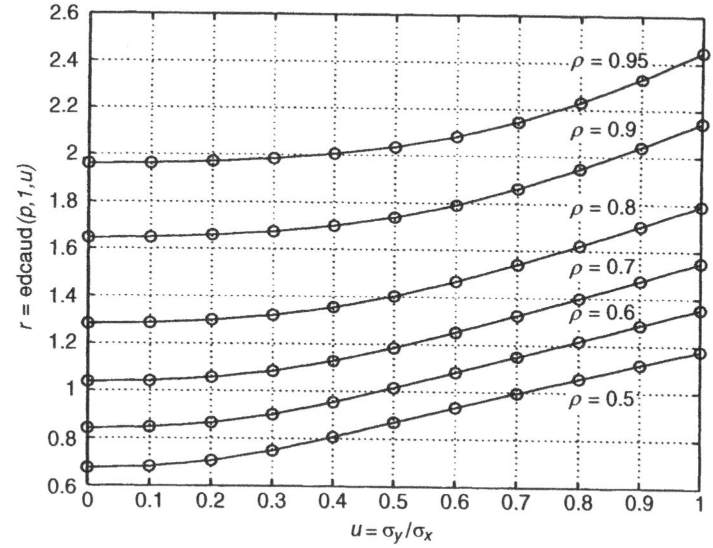

Figure 4.8 Normalized Radius Versus Ellipticity for $\rho = 0$ and Various p Values.

We conclude this section with an example of finding the radius of a 50% probable error circle.

EXAMPLE 4.6

Suppose a certain position error has an elliptical normal distribution with $\sigma_x = 2$ m, $\sigma_y = 3$ m, and that errors in the x and y channels are uncorrelated. Can you find the radius of a circle such that 50% of the time a measured position error will fall inside this circle?

```
R = edcaui(0.5, 2, 3) = 2.9264
```

Thus, we have $R_{50\%} = 2.9264$ m.

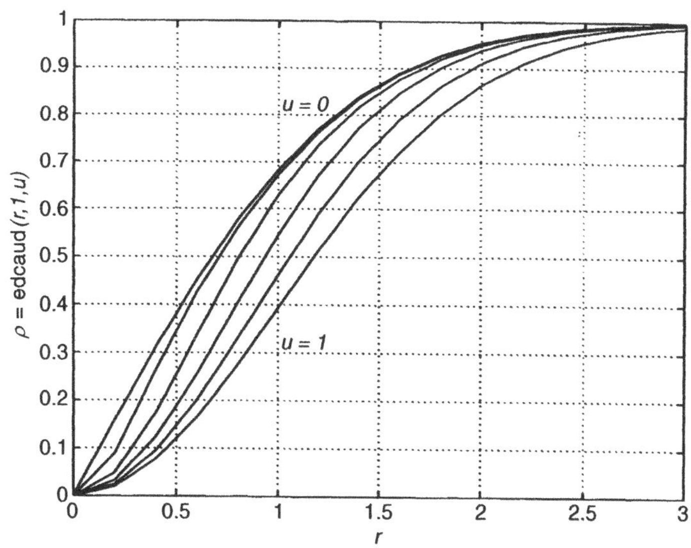

Figure 4.9 Probability Versus Normalized Radius for $u = 0 : 0.2 : 1$ and $\rho = 0$.

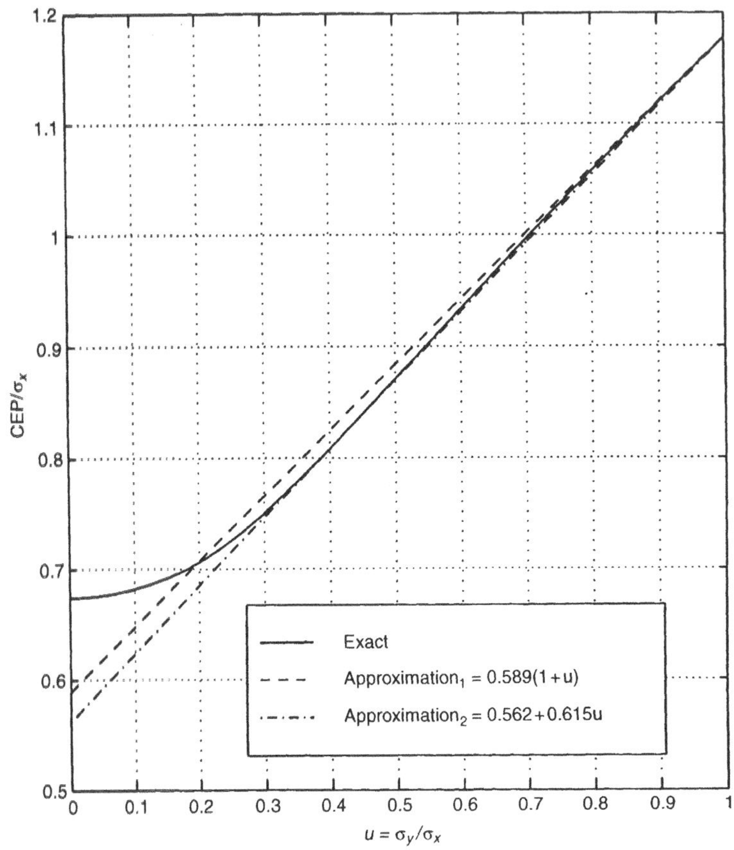

Figure 4.10 CEP/σ_x versus u: Exact and Two Approximations.

4.7 CASE 4, $A = A_c$, $\rho \in [-1, 1]$, $u \in [0, 1]$

Under these assumptions, Equation (4.2) becomes

$$p = \iint_{A_c} \frac{1}{2\pi\sigma_x \sigma_y \sqrt{1-\rho^2}} \exp\left\{ \frac{-1}{2(1-\rho^2)} \left[\frac{x^2}{\sigma_x^2} - \frac{2\rho xy}{\sigma_x \sigma_y} + \frac{y^2}{\sigma_y^2} \right] \right\} dx\, dy \tag{4.20}$$

with the area of integration being the interior of a circle

$$x^2 + y^2 \leq R^2$$

Let

$$Ax^2 + Bxy + Cy^2 = \frac{1}{(1-\rho^2)} \left[\frac{x^2}{\sigma_x^2} - \frac{2\rho xy}{\sigma_x \sigma_y} + \frac{y^2}{\sigma_y^2} \right] \tag{4.21}$$

We can eliminate the second term (Bxy) in the above quadratic expression by a suitable angular rotation θ,

$$\theta = \frac{1}{2}\tan^{-1}\left[\frac{B}{A-C}\right]$$

Substituting the rotation transformation pair

$$x = \xi \cos\theta - \eta \sin\theta \tag{4.22}$$

$$y = \xi \sin\theta + \eta \cos\theta \tag{4.23}$$

into Equation (4.21), we obtain

$$Ax^2 + Bxy + Cy^2 = A_1\xi^2 + C_1\eta^2 \tag{4.24}$$

where the new coefficients A_1, C_1 are related to the old ones (A, B, C) via

$$A_1 = A\cos^2\theta + B\cos\theta\sin\theta + C\sin^2\theta \tag{4.25}$$

$$C_1 = A\sin^2\theta - B\cos\theta\sin\theta + C\cos^2\theta \tag{4.26}$$

This is equivalent to using a similarity transformation on the quadratic form

$$Ax^2 + Bxy + Cy^2 = [x \quad y]\begin{bmatrix} A & B/2 \\ B/2 & A \end{bmatrix}\begin{bmatrix} x \\ y \end{bmatrix}$$

to obtain

$$\begin{aligned} Ax^2 + Bxy + Cy^2 &= [x \quad y]\begin{bmatrix} q_{11} & q_{21} \\ q_{12} & q_{22} \end{bmatrix}\begin{bmatrix} A_1 & 0 \\ 0 & C_1 \end{bmatrix}\begin{bmatrix} q_{11} & q_{12} \\ q_{21} & q_{22} \end{bmatrix}\begin{bmatrix} x \\ y \end{bmatrix} \\ &= A_1\xi^2 + C_1\eta^2 = [\xi \quad \eta]\begin{bmatrix} A_1 & 0 \\ 0 & C_1 \end{bmatrix}\begin{bmatrix} \xi \\ \eta \end{bmatrix} \end{aligned}$$

where A_1, C_1 are the eigenvalues of the matrix

$$\begin{bmatrix} A & B/2 \\ B/2 & A \end{bmatrix}$$

and the two corresponding eigenvectors

$$\begin{bmatrix} q_{11} \\ q_{21} \end{bmatrix} \quad \text{and} \quad \begin{bmatrix} q_{12} \\ q_{22} \end{bmatrix}$$

are used in the transformation from (x, y) to (ξ, η)

$$\begin{bmatrix} \xi \\ \eta \end{bmatrix} = \begin{bmatrix} q_{11} & q_{12} \\ q_{21} & q_{22} \end{bmatrix}\begin{bmatrix} x \\ y \end{bmatrix}$$

The circular boundary of the integration area remains a circle in the new (ξ, η)-coordinate system,

$$\xi^2 + \eta^2 = R^2$$

Note that the problem is now expressed in (ξ, η) and shows no dependency on ρ. Thus, the problem can be solved by the method used in case 3 as

$$p = \iint_{A'_c} \frac{1}{2\pi\sigma_\xi \sigma_\eta} \exp\left\{\frac{-1}{2}\left[\frac{\xi^2}{\sigma_\xi^2} + \frac{\eta^2}{\sigma_\eta^2}\right]\right\} d\xi\, d\eta$$

where $\sigma_\xi^2 = 1/A_1$, $\sigma_\eta^2 = 1/C_1$ and the area of integration A'_c is

$$\xi^2 + \eta^2 \leq R^2$$

Direct problems are solved by **edcacd.m**, with the MATLAB command $p = edcacd(R, \sigma_x, \sigma_y, \rho)$, whereas inverse problems are solved by **edcaci.m** via the command $R = edcaci(p, \sigma_x, \sigma_y, \rho)$. In edcacd.m, the MATLAB program **abc2ac.m** is utilized for the needed rotation transformation before calling edcaud.m.

Figure 4.11 shows the variation of the probability p as u is increased from 0 to 1, for the correlated case corresponding to $\rho = 0.3$. The curve on the left ($R =$ 1 DRMS) indicates that p decreases as u increases; the one on the right ($R = 2$

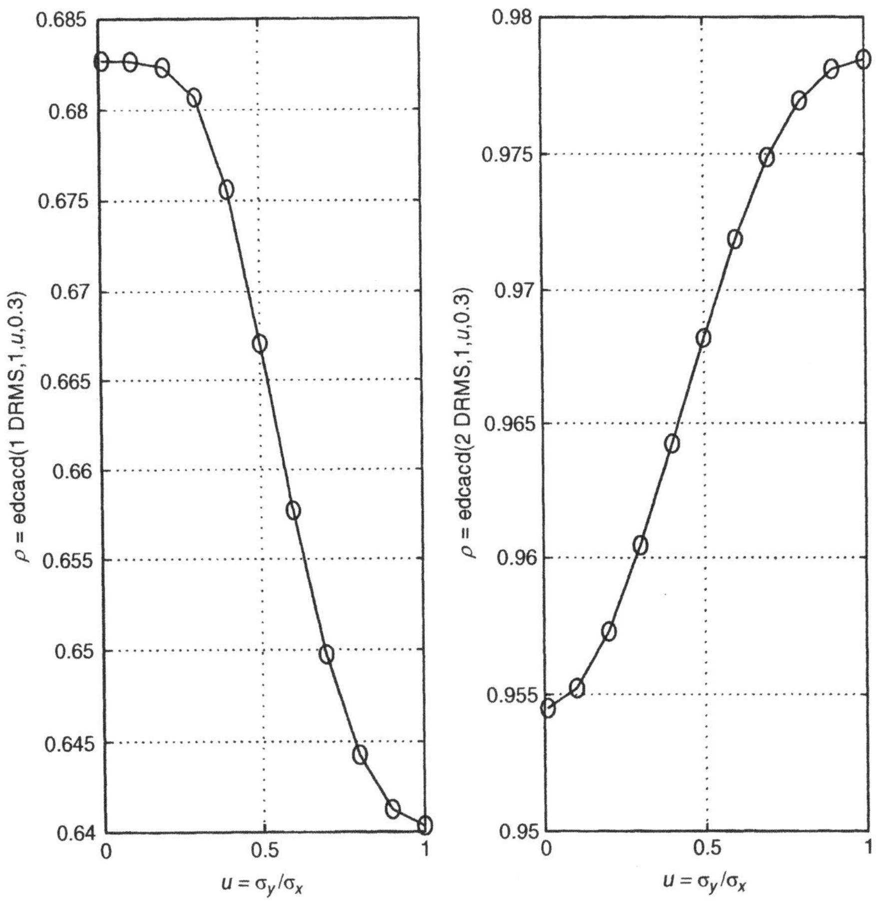

Figure 4.11 Probability Versus Ellipticity for $\rho = 0.3$ and $R = 1$ DRMS and $R = 2$ DRMS.

DRMS) displays a different behavior: p increases as u increases. Recall that as u varies, 1 DRMS = $\sqrt{1+u^2}\,\sigma_x$ does not remain constant. We see similarity in the behavior of these curves and those of Figure 4.7.

Notice that when $\rho = 0$, case 4 reduces to case 3; and when both $\rho = 0$ and $u = 1$, case 4 reduces to case 2.

4.7.1 Limiting Situation: $|\rho| \to 1$

When $\rho = 1$, the two random variables **x**, **y** are perfectly correlated by the linear equation

$$y = x \tan\theta = x(\sigma_y/\sigma_x)$$

Each of the random points will fall on this line; thus, we have a one-dimensional error $\mathbf{t} \sim N(0, \sigma_t^2)$, with $\sigma_t^2 = (1 + u^2)\sigma_x^2$, where **t** represents the random radial distance from (0, 0) to (x, y), $t^2 = x^2 + y^2$.

Our double integral problem becomes a single integral

$$p = \int_{-R}^{R} \frac{1}{\sqrt{2\pi}\sigma_t} \exp\left[\frac{-t^2}{2\sigma_t^2}\right] dt = \text{erf}\left[\frac{R}{\sqrt{2}\sigma_t}\right]$$

Figure 4.12 illustrates the situation.

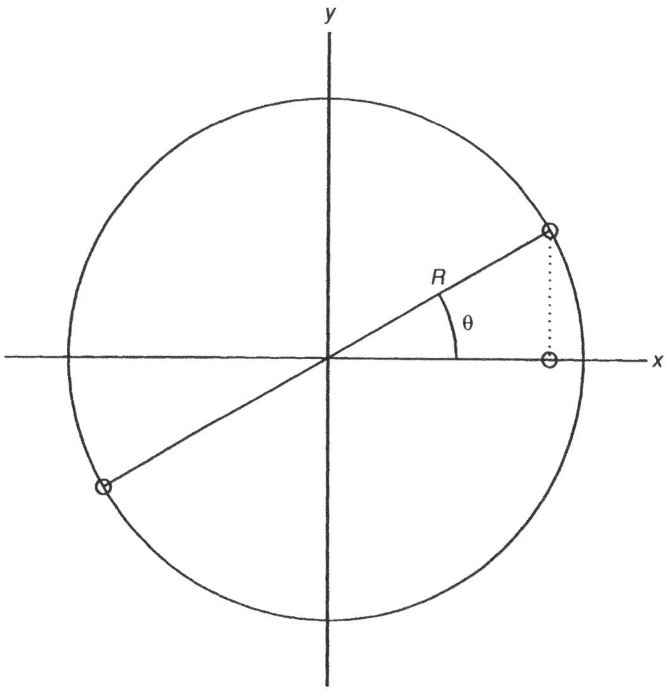

Figure 4.12 Geometry for Limiting Situation: $\rho = 1$.

If $\rho = -1$, the two random variables **x**, **y** are also perfectly correlated; they have the following linear relationship

$$y = -x \tan\theta = -x(\sigma_y/\sigma_x)$$

and all other relationships apply, including Figure 4.12, as for the $\rho = 1$ case.

For fixed $R = 1$, the probability p versus correlation coefficient ρ curves, using the ellipticity u as a parameter, are shown in Figure 4.13. Since the probability is symmetric about $\rho = 0$, the horizontal axis is labeled as $|\rho|$. The probability value increases as $|\rho|$ approaches 1 and as u approaches 0.

4.8 CASE 5, $A = A_e, \quad \rho = 0, \quad u \in [0, 1]$

In this case, the probability in Equation (4.2) becomes

$$p = \iint_A \frac{1}{2\pi\sigma_x\sigma_y} \exp\left\{\frac{-1}{2}\left[\frac{x^2}{\sigma_x^2} + \frac{y^2}{\sigma_y^2}\right]\right\} dx\,dy \tag{4.27}$$

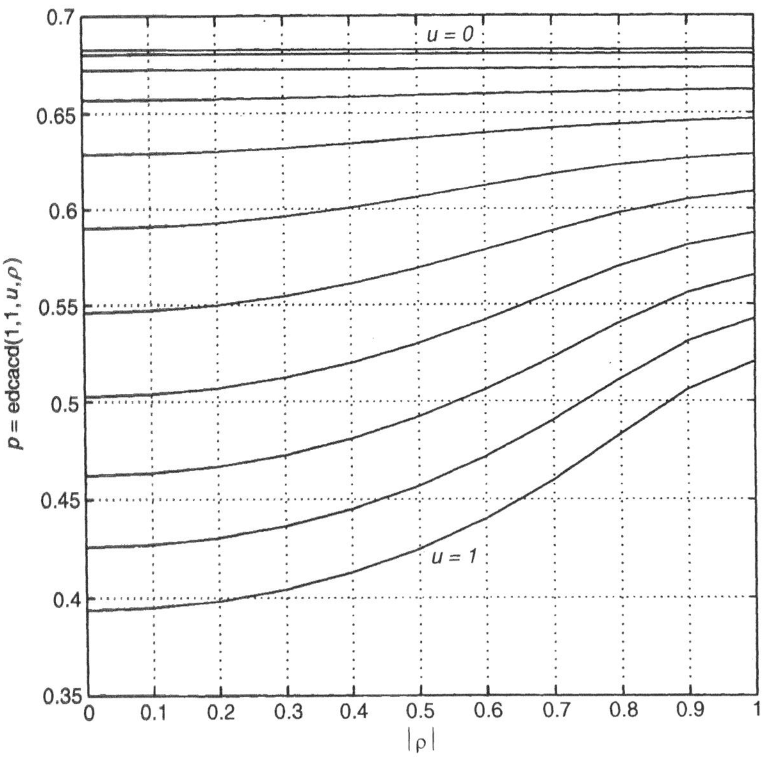

Figure 4.13 Probability Versus Correlation Coefficient for $R = 1$, $u = 0 : 0.1 : 1$.

where the area of interest is an ellipse A_e,

$$\frac{x^2}{a^2} + \frac{y^2}{b^2} \leq 1$$

We solve this problem by using a pair of scaled variables

$$x_1 = x/a \quad \text{and} \quad x_2 = y/b$$

so that Equation (4.27) becomes

$$p = \iint_{A'_c} \frac{1}{2\pi\sigma_1\sigma_2} \exp\left\{\frac{-1}{2}\left[\frac{x_1^2}{\sigma_1^2} + \frac{x_2^2}{\sigma_2^2}\right]\right\} dx_1\, dx_2 \tag{4.28}$$

where $\sigma_1 = \sigma_x/a$, $\sigma_2 = \sigma_y/b$, and the area of integration A'_c is a unit circle in the new coordinates and

$$x_1^2 + x_2^2 \leq 1$$

Once again we have converted the direct problem of Equation (4.27) to one that is solvable by the method used in case 3.

The M-files **edeaud.m** and **edeaui.m** have been developed to solve the direct and inverse problems. The routine edeaud.m includes the needed scaling before calling edcaud.m. The MATLAB commands are $p = edeaud(a, b, \sigma_x, \sigma_y)$ and $[a, b] = edeaui(p, v, \sigma_x, \sigma_y)$.

Notice that when $a = b$, case 5 is simplified to case 3; when $a = b$ and $u = 1$, case 5 is reduced to case 2.

The following example shows how to find a 50%-error ellipse.

EXAMPLE 4.7

If a certain position error has an elliptical normal distribution with $\sigma_x = 2$ m, $\sigma_y = 3$ m, and errors in the x and y channels are uncorrelated, find the semi-major and semi-minor axes (a, b) of an ellipse such that 50% of the time a sample position error will fall inside this ellipse. First, assume $v = (a/b) = 2/3$; next, try $v = (a/b) = 0.8$.

```
[a, b] = edeaui(0.5, 2/3, 2, 3) = [2.3548, 3.5322]
```

Thus, the 50%-error ellipse has a semi-major axis of 3.5322 m in the y direction and a semi-minor axis of 2.3548 m in the x direction. In this case, since $v = 2/3$ coincides with the ellipticity $\sigma_x/\sigma_y = 2/3$, the two ratios a/σ_x and b/σ_y should also be identical; indeed, they are both equal to 1.1774. Why this particular phenomenon occurs (when $a/b = \sigma_x/\sigma_y$) is explained in Subsection 4.8.1 of this chapter.

However, when we set $v = 0.8$ and determine a and b from

```
[a, b] = edeaui(0.5, 0.8, 2, 3) = [2.5871, 3.2338]
```

we can no longer expect the two ratios a/σ_x and b/σ_y to be equal.

Figure 4.14 shows the 50%-error circle and 50%-error ellipse for $p = 0.5$. The values of the parameters used are taken from Examples 4.6 and 4.7; that is, $\sigma_x = 2$, $\sigma_y = 3$, and $p = 0.5$. The semi-major axis and semi-minor axis for the 50% error

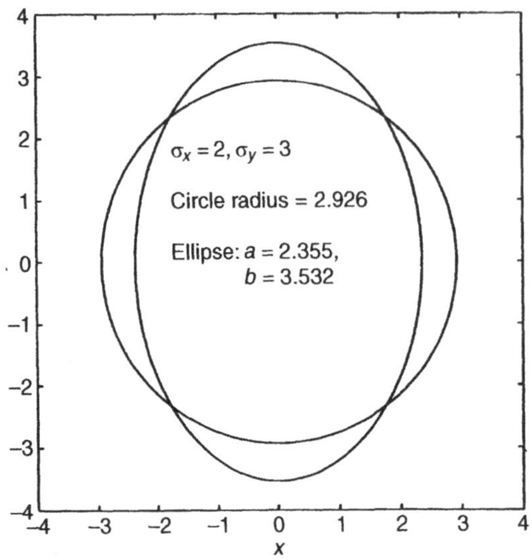

Figure 4.14 p-Error-Circle and p-Error-Ellipse, $p = 50\%$.

ellipse are $a = 2.3548$ and $b = 3.5322$, respectively, whereas the radius for the 50% error-circle is $R = 2.9264$.

It is interesting to note that under the same joint PDF, and for the same probability p, the area associated with the p-error-circle is larger than that of the p-error-ellipse. In Figure 4.14 $A_c = \pi R^2 = 26.9040$, while $A_e = \pi ab = 26.1312$. As the ellipticity u approaches 1, the two areas become closer and eventually coincide. Figure 4.15 shows this relationship between the areas of the 50%-error-circle and 50%-error-ellipse.

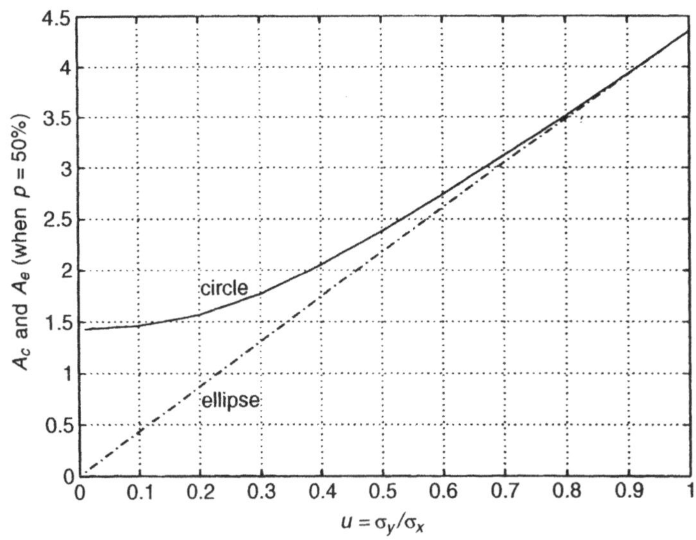

Figure 4.15 Area of 50%-Error-Circle and Area of 50%-Error-Ellipse versus Ellipticity.

4.8.1 Limiting Situation: $a/b = \sigma_x/\sigma_y$

When $a/b = \sigma_x/\sigma_y$, Equation (4.28) can be greatly simplified to

$$p = 1 - \exp(-k^2/2), \quad \text{with } k = a/\sigma_x = b/\sigma_y$$

because $\sigma_1 = \sigma_x/a = \sigma_y/b = \sigma_2$.

This last equation is similar to Equation (4.9). Thus, an elliptical normal PDF over an elliptical region (such that $a/b = \sigma_x/\sigma_y$) can be treated just as the situation of a circular normal PDF over a circular region.

4.9 CASE 6, $A = A_e$, $\rho \in [-1, 1]$, $u \in [0, 1]$

The only difference between the present case and case 4 is that the integration area is now the inside of an ellipse

$$\frac{x^2}{a^2} + \frac{y^2}{b^2} \leq 1$$

instead of a circle $x^2 + y^2 \leq R^2$.

Therefore, if we rescale the variables x and y by

$$x_1 = x/a$$
$$x_2 = y/b$$

then Equation (4.2) becomes

$$p = \iint_{A'_c} \frac{1}{2\pi\sigma_1\sigma_2\sqrt{1-\rho^2}} \exp\left\{\frac{-1}{2(1-\rho^2)}\left[\frac{x_1^2}{\sigma_1^2} - \frac{2\rho x_1 x_2}{\sigma_1 \sigma_2} + \frac{x_2^2}{\sigma_2^2}\right]\right\} dx_1\, dx_2 \tag{4.29}$$

where $\sigma_1 = \sigma_x/a$, $\sigma_2 = \sigma_y/b$, and the area of integration A'_c is a circle in the new (x_1, x_2) coordinates and

$$x_1^2 + x_2^2 \leq 1$$

We can see that this equation is almost identical to Equation (4.20), except radius R in that equation is replaced by 1 here.

The M-files **edeacd.m** and **edeaci.m** have been developed to solve the direct and inverse problems, respectively. The two MATLAB commands are $p = edeacd(a, b, \sigma_x, \sigma_y, \rho)$ and $[a, b] = edeaci(p, v, \sigma_x, \sigma_y, \rho)$.

Notice that when $\rho = 0$, case 6 is specialized to case 5; when $a = b$, case 6 simplifies to case 4; when $\rho = 0$ and $a = b$, case 6 collapses to case 3; when $\rho = 0$, $a = b$, and $u = 1$, case 6 reduces to case 2. This is the most general case among cases 2 through 6.

For the case of $a = 2$ and $b = 1.5$, probability p versus correlation coefficient ρ curves, using ellipticity u as a parameter, are shown in Figure 4.16. Since the probability is symmetric about $\rho = 0$, the horizontal axis is labeled as $|\rho|$. The probability value increases as u approaches 0.

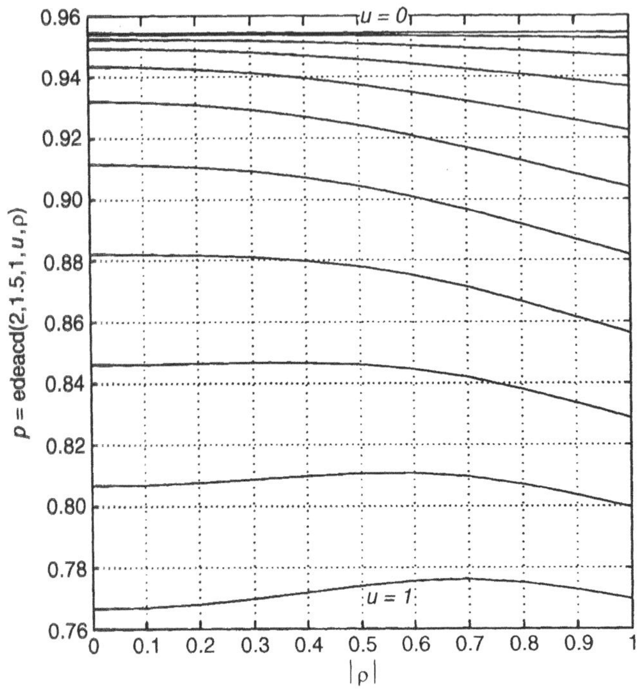

Figure 4.16 Probability Versus ρ for $a = 2$, $b = 1.5$, and $u = 0 : 0.1 : 1$.

4.9.1 Limiting Situation: $|\rho| \to 1$

When $\rho = 1$, the two random variables **x**, **y** are perfectly correlated by the linear equation

$$y = xu = x \tan\theta = x(\sigma_y/\sigma_x)$$

Each of the random points will fall on this line. Thus, we have a one-dimensional error $\mathbf{t} \sim N(0, \sigma_t^2)$, with $\sigma_t^2 = (1 + u^2)\sigma_x^2$, where **t** represents the radial distance from (0, 0) to (x, y).

Our double integral problem becomes a single integral

$$p = \int_{-t_0}^{t_0} \frac{1}{\sqrt{2\pi}\sigma_t} \exp\left[\frac{-t^2}{2\sigma_t^2}\right] dt = \operatorname{erf}\left[\frac{t_0}{\sqrt{2}\sigma_t}\right]$$

where the integration limit (t_0) is found as follows:

First, solve

$$y = xu = x \tan\theta \tag{4.30}$$

$$\frac{x^2}{a^2} + \frac{y^2}{b^2} = 1 \tag{4.31}$$

and obtain

$$x^2 = \frac{a^2 b^2}{a^2 u^2 + b^2}$$

$$y^2 = \frac{a^2 b^2 u^2}{a^2 u^2 + b^2}$$

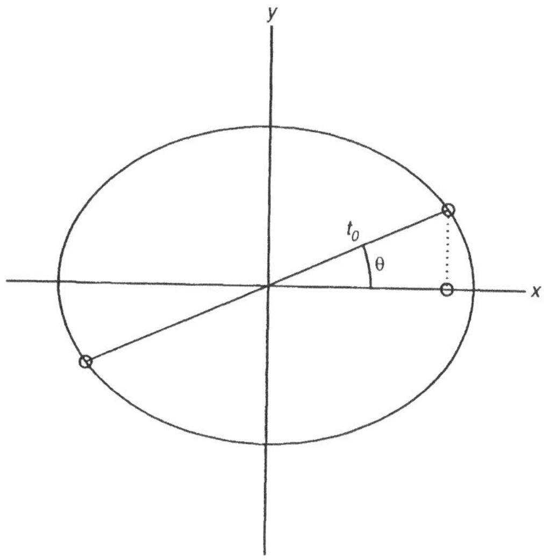

Figure 4.17 Geometry for Limiting Situation: $\rho = 1$.

Thus,

$$t_0^2 = x^2 + y^2 = \frac{a^2b^2(1+u^2)}{a^2u^2+b^2}$$

Figure 4.17 illustrates the situation. If $\rho = -1$, the only difference is in the line

$$y = -xu = -x\tan\theta = -x(\sigma_y/\sigma_x)$$

4.10 PDF OF THE ANGULAR POSITION OF A RANDOM POINT

In the preceding sections we have studied the probability that a random radial error falls in some particular region surrounding the origin. Here we look at the probability distribution for angular position of a random point on the plane.

Assume that the rectangular coordinates $(\mathbf{x}, \mathbf{y})$ of a random point are independent random variables with probability density functions

$$\mathbf{x} \sim N(\mu_x, \sigma_x^2) \tag{4.32}$$

$$\mathbf{y} \sim N(\mu_y, \sigma_y^2) \tag{4.33}$$

What is the probability density function (PDF) of the polar angle of a random point in the plane relative to the origin (0, 0)?

The problem can be solved by first converting the two-dimensional joint PDF in rectangular coordinates (x, y) into polar coordinates (r, θ), and then integrating with respect to r from 0 to ∞ to find $h(\theta)$, the PDF of the random variable $\boldsymbol{\theta}$.

This PDF obviously is a function of θ, but it also depends on four other quantities: $\mu_x, \sigma_x, \mu_y, \sigma_y$.

The joint probability density function for the two random variables (**x**, **y**) is

$$f(x, y) = \frac{1}{2\pi\sigma_x \sigma_y} \exp\left\{-\frac{1}{2}\left[\frac{(x-\mu_x)^2}{\sigma_x^2} + \frac{(y-\mu_y)^2}{\sigma_y^2}\right]\right\}$$

Using the Jacobian operator

$$\frac{\partial(x, y)}{\partial(r, \theta)} = \begin{vmatrix} \partial x/\partial r & \partial x/\partial\theta \\ \partial y/\partial r & \partial y/\partial\theta \end{vmatrix} = \begin{vmatrix} \cos\theta & -r\sin\theta \\ \sin\theta & r\cos\theta \end{vmatrix} = r$$

we obtain the joint PDF $g(r, \theta)$ for the two new random variables (**r**, **θ**)

$$g(r, \theta) = \frac{\partial(x, y)}{\partial(r, \theta)} f(x, y) = \frac{r}{2\pi\sigma_x\sigma_y} \exp\left\{-\frac{1}{2}\left[\frac{(r\cos\theta-\mu_x)^2}{\sigma_x^2} + \frac{(r\sin\theta-\mu_y)^2}{\sigma_y^2}\right]\right\}$$

Defining

$$A = \frac{\cos^2\theta}{\sigma_x^2} + \frac{\sin^2\theta}{\sigma_y^2} \tag{4.34}$$

$$B = -2\left[\frac{\mu_x\cos\theta}{\sigma_x^2} + \frac{\mu_y\sin\theta}{\sigma_y^2}\right] \tag{4.35}$$

$$C = \frac{\mu_x^2}{\sigma_x^2} + \frac{\mu_y^2}{\sigma_y^2} \tag{4.36}$$

we simplify $g(r, \theta)$ to

$$g(r, \theta) = \frac{r}{2\pi\sigma_x\sigma_y} \exp\left\{-\frac{1}{2}\left[A\left(r+\frac{B}{2A}\right)^2 + \left(C-\frac{B^2}{4A}\right)\right]\right\} \tag{4.37}$$

The one-dimensional PDF, $h(\theta)$, for the random variable **θ** is obtained by integrating Equation (4.37) with respect to r from 0 to ∞,

$$h(\theta) = \int_0^\infty \frac{r}{2\pi\sigma_x\sigma_y} \exp\left\{-\frac{1}{2}\left[A\left(r+\frac{B}{2A}\right)^2 + \left(C-\frac{B^2}{4A}\right)\right]\right\} dr \tag{4.38}$$

After a sequence of algebraic operations and simplifications, we have

$$h(\theta) = \frac{1}{2A\pi\sigma_x\sigma_y} \exp(-C/2)\left[1 - w\sqrt{\pi}\exp(w^2)\operatorname{erfc}(w)\right] \tag{4.39}$$

where erfc, the complementary error function, and w are defined as

$$\operatorname{erfc}(t) = \frac{2}{\sqrt{\pi}} \int_t^\infty \exp(-u^2)\, du \tag{4.40}$$

$$w = \frac{B}{\sqrt{8A}} \tag{4.41}$$

The M-file **pdft.m** performs the operations required to compute $h(\theta)$ when μ_x, μ_y, σ_x, and σ_y are given. The command form in MATLAB is $y = pdft(\theta, \mu_x, \mu_y, \sigma_x, \sigma_y)$. Figure 4.18 shows $h(\theta)$ versus θ for $\mu_x = 5$, $\mu_y = 3$, $\sigma_x = 0.4$, and $\sigma_y = 0.25$.

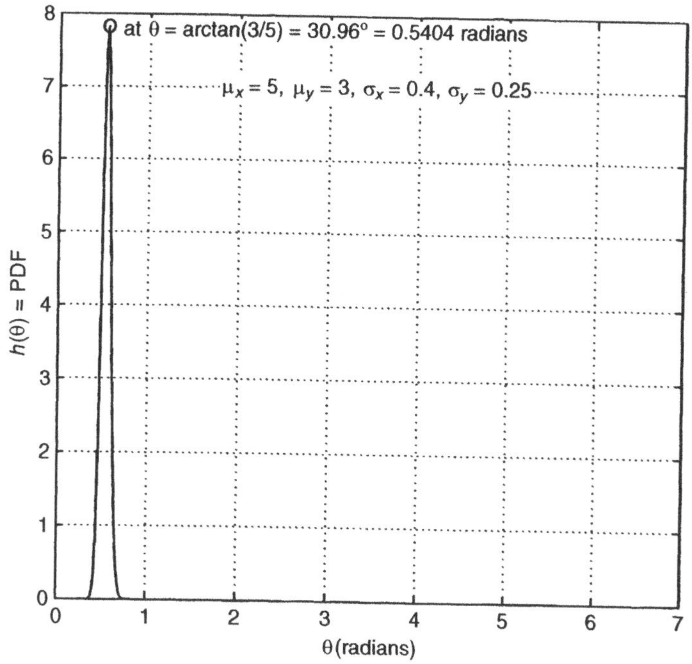

Figure 4.18 Probability Density Function of Polar Angle θ.

When $\mu_x = 0,\ \mu_y = 0$, and $\sigma_x = \sigma_y = \sigma,\ A,\ B,\ C,\ g(r, \theta)$ become

$$A = \frac{1}{\sigma^2}$$

$$B = C = w = 0$$

$$g(r, \theta) = \frac{r}{2\pi\sigma^2} \exp\left[-\frac{r^2}{2\sigma^2}\right]$$

and the PDF in Equation (4.39) reduces to a constant over the interval $[0, 2\pi]$ as

$$h(\theta) = \frac{1}{2\pi}, \quad \text{for } \theta \in [0, 2\pi] \tag{4.42}$$

In this case, the random angular position (measured in radians) has a simple, uniform distribution over $[0, 2\pi]$.

4.11 SUMMARY OF M-FILES

The M-files used or generated in this chapter are summarized in Table 4.4. Recall that the true length R and the normalized length r are related by

$$r = R/\max(\sigma_x, \sigma_y)$$

TABLE 4.4 M-files Used in Chapter 4

File Name	MATLAB Command
nf2.m	$p = nf2(r)$
cf1.m	$y = cf1(r)$
cf2.m	$p = cf2(r)$
cf3.m	$r = cf3(p)$
gf2.m	$p = gf2(r, u)$
simprule.m	$s = simprule(y, h)$
r2p2d.m	$r2p2d(k)$
p2r2d.m	$p2r2d(p)$
edraud.m	$p = edraud(a, b, \sigma_x, \sigma_y)$
edraui.m	$[a, b] = edraui(p, v, \sigma_x, \sigma_y)$
edca.m	$p = edca(r, u)$
cdcaud.m	$p = cdcaud(R, \sigma_x)$
cdcaui.m	$R = cdcaui(p, \sigma_x)$
edcaud.m	$p = edcaud(R, \sigma_x, \sigma_y)$
edcaui.m	$R = edcaui(p, \sigma_x, \sigma_y)$
edcacd.m	$p = edcacd(R, \sigma_x, \sigma_y, \rho)$
edcaci.m	$R = edcaci(p, \sigma_x, \sigma_y, \rho)$
edeaud.m	$p = edeaud(a, b, \sigma_x, \sigma_y)$
edeaui.m	$[a, b] = edeaui(p, v, \sigma_x, \sigma_y)$
edeacd.m	$p = edeaud(a, b, \sigma_x, \sigma_y, \rho)$
edeaci.m	$[a, b] = edeaui(p, v, \sigma_x, \sigma_y, \rho)$
abc2ac.m	$[a1, c1] = abc2ac(a, b, c)$
pdft.m	$y = pdft(\theta, \mu_x, \mu_y, \sigma_x, \sigma_y)$

The M-file **edca.m** is invoked in the program edcaud.m.

Figure 4.19 describes the relationships between the major M-files developed in this chapter.

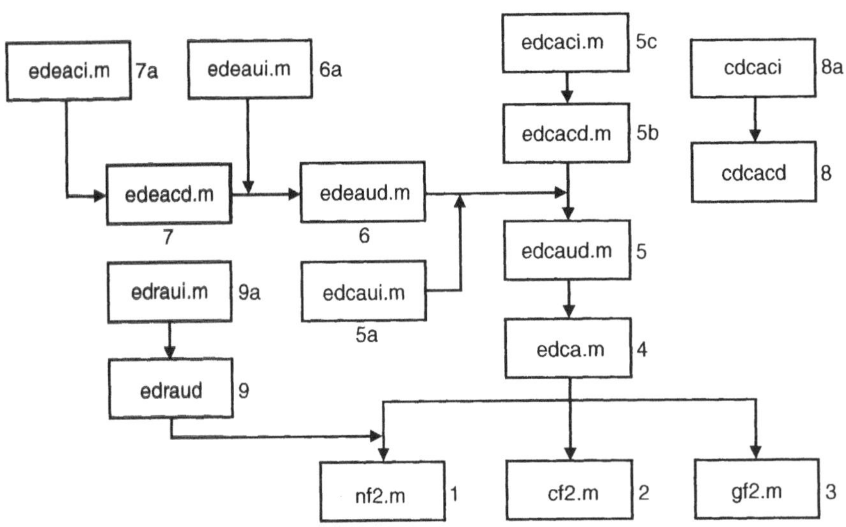

Figure 4.19 Relationships Between Major M-files.

5

Three-Dimensional Error Analysis

Errors in the measurement of distance in three-dimensional space are treated in this chapter. This type of error is the result of a root sum square process of three one-dimensional random errors in three different channels.

5.1 THREE-DIMENSIONAL NORMAL DISTRIBUTION

In three-dimensional error analysis, the most general **trivariate normal** joint probability density function of random variables (**x**, **y**, **z**) is

$$
\begin{aligned}
g(x, y, z) = {} & \frac{1}{\sqrt{(2\pi)^3(1 + 2\rho_{xy}\rho_{yz}\rho_{xz} - \rho_{xy}^2 - \rho_{yz}^2 - \rho_{xz}^2)}\,\sigma_x\sigma_y\sigma_z} \\
& \times \exp\left\{\frac{-1}{2(1 + 2\rho_{xy}\rho_{yz}\rho_{xz} - \rho_{xy}^2 - \rho_{yz}^2 - \rho_{xz}^2)}\right. \\
& \times \left[\frac{(1-\rho_{yz}^2)(x-\mu_x)^2}{\sigma_x^2} + \frac{(1-\rho_{xz}^2)(y-\mu_y)^2}{\sigma_y^2} + \frac{(1-\rho_{xy}^2)(z-\mu_z)^2}{\sigma_z^2}\right. \\
& - \frac{2(\rho_{xy} - \rho_{xz}\rho_{yz})(x-\mu_x)(y-\mu_y)}{\sigma_x\sigma_y} - \frac{2(\rho_{yz} - \rho_{xy}\rho_{xz})(y-\mu_y)(z-\mu_z)}{\sigma_y\sigma_z} \\
& \left.\left. - \frac{2(\rho_{xz} - \rho_{xy}\rho_{yz})(x-\mu_x)(z-\mu_z)}{\sigma_x\sigma_z}\right]\right\}
\end{aligned}
$$

where μ_x and σ_x are the mean and standard deviation for random variable **x**; μ_y and σ_y are the mean and standard deviation for random variable **y**; μ_z and σ_z are the mean and standard deviation for random variable **z**; and ρ_{xy}, ρ_{yz}, ρ_{xz} are the **correlation coefficients** between the xy, yz, xz channels, respectively.

After the means μ_x, μ_y, and μ_z are removed, the joint normal PDF is simplified to

$$f(x, y, z) = \frac{1}{\sqrt{(2\pi)^3(1 + 2\rho_{xy}\rho_{yz}\rho_{xz} - \rho_{xy}^2 - \rho_{yz}^2 - \rho_{xz}^2)}\,\sigma_x\sigma_y\sigma_z}$$
$$\times \exp\left\{\frac{-1}{2(1 + 2\rho_{xy}\rho_{yz}\rho_{xz} - \rho_{xy}^2 - \rho_{yz}^2 - \rho_{xz}^2)}\right.$$
$$\times\left[\frac{(1-\rho_{yz}^2)x^2}{\sigma_x^2} + \frac{(1-\rho_{xz}^2)y^2}{\sigma_y^2} + \frac{(1-\rho_{xy}^2)z^2}{\sigma_z^2} - \frac{2(\rho_{xy} - \rho_{xz}\rho_{yz})xy}{\sigma_x\sigma_y}\right.$$
$$\left.\left. - \frac{2(\rho_{yz} - \rho_{xy}\rho_{xz})yz}{\sigma_y\sigma_z} - \frac{2(\rho_{xz} - \rho_{xy}\rho_{yz})xz}{\sigma_x\sigma_z}\right]\right\} \tag{5.1}$$

We assume, without loss of generality, that $\sigma_x \geq \sigma_y \geq \sigma_z$ and use u and v to represent the ratios σ_y/σ_x and σ_z/σ_x, respectively. Since $\sigma_x \geq \sigma_y \geq \sigma_z$, then $1 \geq u \geq v \geq 0$. The two ratios

$$u = \sigma_y/\sigma_x$$
$$v = \sigma_z/\sigma_x$$

will be called the **first and second ellipticities**, respectively.

When $\sigma_x = \sigma_y = \sigma_z = \sigma$ (or $u = v = 1$) and $\rho_{xy} = \rho_{yz} = \rho_{xz} = 0$, the joint PDF becomes

$$f(x, y, z) = \frac{1}{\sqrt{(2\pi)^3\sigma^3}}\exp\left[\frac{-1}{2\sigma^2}(x^2 + y^2 + z^2)\right] \tag{5.2}$$

This PDF is called the **spherical normal distribution** since each volume of equal probability density has the shape of a sphere $x^2 + y^2 + z^2 = k\sigma^2$.

All other values for the σ's and ρ's lead to an **ellipsoidal normal distribution** since each volume of equal probability density has the shape of an ellipsoid.

5.2 DIRECT AND INVERSE PROBLEMS

The goal of the direct problem in three-dimensional error analysis is to evaluate the probability p via the triple integral of $f(x, y, z)$, shown in Equation (5.1), over a given volume V

$$p = \iiint_V f(x, y, z)\,dx\,dy\,dz \tag{5.3}$$

where the integration volume V can be one of three types

Box $V = V_b$: $|x| \leq a,\ |y| \leq b,\ |z| \leq c$
Sphere $V = V_s$: $x^2 + y^2 + z^2 \leq R^2$
Ellipsoid $V = V_e$: $(x/a)^2 + (y/b)^2 + (z/c)^2 \leq 1$

For inverse problems we start with a specified probability p_0 and then proceed to find the dimensions of the integration volume—that is, the width, length, and height of a box; the radius of a sphere; or the three semi-axes of an ellipsoid. The solution is found via a repetitive trial-and-comparison procedure:

(1) Assume an initial dimension(s) for V; choose a reasonably small tolerance limit δ.
(2) Carry out the triple integration in Equation (5.3) to obtain p.
(3) Compare p with p_0.
(4) If $|p - p_0| < \delta$, accurate dimensions of V have been found, stop.
(5) If $p - p_0 > \delta$, decrease the dimensions of V, then go back to (2).
(6) If $p_0 - p > \delta$, increase the dimensions of V, then go back to (2).

Therefore, both the direct problem and the inverse problem require an efficient way to compute the triple integral in Equation (5.3).

Depending on the size of the ellipticities u, v, the shape of the integration volume V, and the value of the correlation coefficients ρ_{ij}, there are six cases to be investigated:

- Case 1, $V = V_b$, $\rho_{xy} = \rho_{yz} = \rho_{xz} = 0$, $u, v \in [0, 1]$
- Case 2, $V = V_s$, $\rho_{xy} = \rho_{yz} = \rho_{xz} = 0$, $u = v = 1$
- Case 3, $V = V_s$, $\rho_{xy} = \rho_{yz} = \rho_{xz} = 0$, $u, v \in [0, 1]$
- Case 4, $V = V_s$, $\rho_{xy}, \rho_{yz}, \rho_{xz} \in [-1, 1]$, $u, v \in [0, 1]$
- Case 5, $V = V_e$, $\rho_{xy} = \rho_{yz} = \rho_{xz} = 0$, $u, v \in [0, 1]$
- Case 6, $V = V_e$, $\rho_{xy}, \rho_{yz}, \rho_{xz} \in [-1, 1]$, $u, v \in [0, 1]$

In order to easily refer to these cases without lengthy description, we rely on an abbreviation system using six characters (XXYYZP). The meaning of the six-character string is as follows:

XX: (ED, SD)	(Ellipsoidal Distribution, Spherical Distribution)
YY: (BV, SV, EV)	(Box, Sphere, Ellipsoidal Volume)
Z: (U, C)	(Uncorrelated, Correlated)
P: (D, I)	(Direct problem, Inverse problem)

Thus, the direct and inverse problems to be considered are:

Case 1	(EDBVUD), (EDBVUI)
Case 2	(SDSVUD), (SDSVUI)
Case 3	(EDSVUD), (EDSVUI)
Case 4	(EDSVCD), (EDSVCI)
Case 5	(EDEVUD), (EDEVUI)
Case 6	(EDEVCD), (EDEVCI)

These cases are studied in Sections 5.3 through 5.9. When naming M-files developed in this chapter, we will also follow the same abbreviation convention. Cases 2 and 3 have been treated in [**9**].

5.3 CASE 1, $V = V_b, \quad \rho_{xy} = \rho_{yz} = \rho_{xz} = 0, \quad u, v \in [0, 1]$

Assuming that the volume of integration is a box and that the three channels are uncorrelated, we can write the triple integral in Equation (5.3) as

$$p = \int_{-c}^{c}\int_{-b}^{b}\int_{-a}^{a} \frac{1}{2\pi\sigma_x\sigma_y\sigma_z} \exp\left\{\frac{-1}{2}\left[\frac{x^2}{\sigma_x^2}+\frac{y^2}{\sigma_y^2}+\frac{z^2}{\sigma_z^2}\right]\right\} dx\, dy\, dz \tag{5.4}$$

which is separable and can be simplified to

$$p = \operatorname{erf}\left[\frac{a}{\sqrt{2}\,\sigma_x}\right] \cdot \operatorname{erf}\left[\frac{b}{\sqrt{2}\,\sigma_y}\right] \cdot \operatorname{erf}\left[\frac{c}{\sqrt{2}\,\sigma_z}\right] \tag{5.5}$$

Even though six parameters a, b, c, σ_x, σ_y, σ_z appear in Equation (5.5), only three normalized variables $a_0 = a/\sigma_x$, $b_0 = b/\sigma_y$, and $c_0 = c/\sigma_z$ are needed to determine the probability p. The direct problem can be solved either by look-up in Table 5.1 (which is identical to Table 4.1; see instructions on page 39) containing the values of $\operatorname{erf}(R_0/\sqrt{2})$, or by using the M-file **edbvud.m**. The MATLAB command to use this file is $p = edbvud([a, b, c], [\sigma_x, \sigma_y, \sigma_z])$. An example follows.

TABLE 5.1 Values of $\operatorname{erf}(R_0/\sqrt{2})$

$R_0 = i + d$	i			
d	0	1	2	3
0.0	0.0000	0.6827	0.9545	0.9973
0.1	0.0797	0.7287	0.9643	0.9981
0.2	0.1585	0.7699	0.9722	0.9986
0.3	0.2358	0.8064	0.9786	0.9990
0.4	0.3108	0.8385	0.9836	0.9993
0.5	0.3829	0.8664	0.9876	0.9995
0.6	0.4515	0.8904	0.9907	0.9997
0.7	0.5161	0.9109	0.9931	0.9998
0.8	0.5763	0.9281	0.9949	0.9999
0.9	0.6319	0.9426	0.9963	0.9999

EXAMPLE 5.1

If a position error has an ellipsoidal normal PDF with $\sigma_x = 2$ nmi, $\sigma_y = 1$ nmi, $\sigma_z = 0.25$ nmi, and errors in the x, y, z channels are uncorrelated, what is the probability that a measured position error will fall within a box bounded by $|x| \le a = 1.4$ nmi, $|y| \le b = 1$ nmi, and $|z| \le c = 0.5$ nmi?

The probability can be found from Table 5.1 with $a_0 = (1.4/2) = 0.7$, $b_0 = 1/1 = 1$, and $c_0 = (0.5/0.25) = 2$. Hence,

$$p = \text{erf}(a_0/\sqrt{2})\text{erf}(b_0/\sqrt{2})\text{erf}(c_0/\sqrt{2})$$
$$= 0.5161 \times 0.6827 \times 0.9545 = 0.3363,$$

or we can enter either of the two MATLAB commands:

```
p = nf2(0.7)*nf2(1)*nf2(2)
p = edbvud([1.4, 1, 0.5], [2, 1, 0.25])
```

and obtain the same result $p = 0.3363$.

Given σ_x, σ_y, σ_z, p, there are many (a, b, c) triplets that satisfy Equation (5.5). The inverse problem could have multiple solutions of (a, b, c) if one does not specify the two ratios $k_1 = a/b$, $k_2 = b/c$. Notice that unlike the ratios u, v, which are bounded by 0 and 1, the ratios k_1, k_2 can have any value in $[0, \infty)$. Solutions for inverse problems are obtainable from **edbvui.m**. The MATLAB command is $[a, b, c] = edbvui(p, k_1, k_2, [\sigma_x, \sigma_y, \sigma_z])$. Let us look at another example.

EXAMPLE 5.2

Suppose that the joint PDF of a certain position error is ellipsoidal normal with $\sigma_x = 1.5$ nmi, $\sigma_y = 0.6$ nmi, $\sigma_z = 0.5$ nmi, and that errors in the x, y, z channels are uncorrelated. Determine the length a along the x-axis, width b along the y-axis, and height c along the z-axis when the specified ratios are $k_1 = (a/b) = 2$, $k_2 = (b/c) = 1$, and the probability for a sample position error to fall inside this $2a \times 2b \times 2c$ box is 90%.

After entering the MATLAB command

```
[a,b,c] = edbvui(0.9, 2, 1,[1.5,0.6,0.5])
```

we obtain $a = 2.7049$, $b = 1.3525$, $c = 1.3525$. If, however, $k_1 = (a/b) = 1/2$, $k_2 = (b/c) = 1$, then using

```
[a,b,c] = edbvui(0.9, 0.5, 1,[1.5,0.6,0.5])
```

we obtain $a = 2.4673$, $b = 4.9346$, $c = 4.9346$, instead. This demonstrates the necessity of clearly specifying $k_1 = (a/b)$, $k_2 = (b/c)$ for the inverse problem.

5.4 CASE 2, $V = V_s$, $\rho_{xy} = \rho_{yz} = \rho_{xz} = 0$, $u = v = 1$

When the random variables **x**, **y**, and **z** have the same standard deviation $(\sigma_x = \sigma_y = \sigma_z = \sigma)$, the general joint PDF is reduced to a spherical normal distribution

$$f(x, y, z) = \frac{1}{\sqrt{(2\pi)^3}\sigma^3} \exp\left[\frac{-1}{2\sigma^2}(x^2 + y^2 + z^2)\right] \tag{5.6}$$

The triple integral of $f(x, y, z)$ over a spherical volume $x^2 + y^2 + z^2 \leq R^2$ represents the probability of a sample point falling in this region. Equation (5.3) can be transformed to an equivalent form in spherical coordinates

$$p = \int_0^{\pi}\int_0^{2\pi}\int_0^{R} \frac{1}{\sqrt{(2\pi)^3}\sigma^3} \exp\left[\frac{-r^2}{2\sigma^2}\right] r^2 dr \sin(\phi)\, d\phi\, d\theta = \int_0^R h(r, \sigma)\, dr \tag{5.7}$$

with $h(r, \sigma)$ being the **Maxwell distribution function**

$$h(r, \sigma) = \sqrt{\frac{2}{\pi}} \frac{r^2}{\sigma^3} \exp\left[\frac{-r^2}{2\sigma^2}\right], \quad \text{for } r \geq 0 \tag{5.8}$$

The independent variable of this Maxwell probability density function is the random radial error $\mathbf{r} = \sqrt{\mathbf{x}^2 + \mathbf{y}^2 + \mathbf{z}^2}$. The M-file **sf1.m** implements the Maxwell PDF for $\sigma = 1$.

Following the notation convention of Chapter 3 (page 25), we will designate the associated CDF as $H(R/\sigma)$, where

$$H(R/\sigma) = \int_0^R h(t, \sigma)\, dt$$

since we have used $h(r, \sigma)$ as the PDF of $\mathbf{r}$. Thus, carrying out the integration in Equation (5.7), we have

$$\begin{aligned} p = H(R/\sigma) &= \int_0^R \sqrt{\frac{2}{\pi}} \frac{t^2}{\sigma^3} \exp\left[\frac{-t^2}{2\sigma^2}\right] dt \\ &= -\sqrt{\frac{2}{\pi}} \frac{R}{\sigma} \exp\left[\frac{-R^2}{2\sigma^2}\right] + \operatorname{erf}\left(\frac{R}{\sqrt{2}\sigma}\right) \end{aligned} \tag{5.9}$$

Geometrically, this means that the integral of $f(x, y, z)$ over the volume $x^2 + y^2 + z^2 \leq R^2$ is equivalent to the area under the Maxwell PDF, $h(r, \sigma)$, over the interval $[0, R]$. Therefore, in this special case, and this case alone, three-dimensional error analysis can be performed in terms of a single random variable $\mathbf{r}$, the spatial (three-dimensional) radial error.

Equation (5.9) provides the closed-form solution for the direct problem. We can rewrite Equation (5.9) in the following form

$$p = H(r) = H(R/\sigma) - \sqrt{\frac{2}{\pi}} r \exp\left[\frac{-r^2}{2}\right] + \operatorname{erf}\left(\frac{r}{\sqrt{2}}\right) \tag{5.10}$$

Note that $r = R/\sigma$ is the radius normalized with respect to σ. The M-file **sf2.m**, with the MATLAB command form $p = sf2(r)$, can be used to find the probability p.

The inverse problem of finding R from

$$p = -\sqrt{\frac{2}{\pi}} \frac{R}{\sigma} \exp\left[\frac{-R^2}{2\sigma^2}\right] + \operatorname{erf}\left(\frac{R}{\sqrt{2}\sigma}\right)$$

or finding r from

$$p = -\sqrt{\frac{2}{\pi}} r \exp\left[\frac{-r^2}{2}\right] + \operatorname{erf}\left(\frac{r}{\sqrt{2}}\right)$$

with given p, cannot be solved analytically. Numerical iteration is used to obtain the solution. This numerical inversion procedure is implemented in **sf3.m**. The MATLAB command is $r = sf3(p)$. Another command, $r = sf4(p)$, achieves the same goal using the Newton–Raphson method; see Equation (1.6). The corresponding general forms of sf2.m and sf3.m are **sdsvud.m** and **sdsvui.m**, respectively. The MATLAB commands to use these files are $p = sdsvud(R, \sigma_x)$ and $R = sdsvui(p, \sigma_x)$.

5.5 THREE-DIMENSIONAL ERROR MEASURES

Because of its simplicity, the Maxwell distribution will be used to illustrate several three-dimensional error indicators.

5.5.1 Mean Radial Error (MRE)

The mean radial error (μ_r) is defined as

$$\text{MRE} = \mu_r = E[\mathbf{r}] = \int_0^{\infty} rh(r, \sigma)\,dr = 2\sqrt{\frac{2}{\pi}}\sigma \approx 1.5958\sigma \tag{5.11}$$

5.5.2 Root Mean Square Radial Error (RMSR)

The mean square radial (**MSR**) error is similarly defined as

$$\text{MSR} = E[\mathbf{r}^2] = \int_0^{\infty} r^2 h(r, \sigma)\,dr = 3\sigma^2$$

The RMSR is obtained by taking the square root of the MSR

$$\text{RMSR} = \sqrt{E[\mathbf{r}^2]} = \sqrt{3}\,\sigma \approx 1.7321\sigma \tag{5.12}$$

5.5.3 Distance Root Mean Square Error (DRMS)

DRMS is defined as the square root of the sum of all three x, y, z variances as

$$\text{DRMS} = \sqrt{\sigma_x^2 + \sigma_y^2 + \sigma_z^2} = \sqrt{1 + u^2 + v^2}\,\sigma_x \tag{5.13}$$

with $u = \sigma_y/\sigma_x$ and $v = \sigma_z/\sigma_x$. When $\sigma_x = \sigma_y = \sigma_z = \sigma$,

$$\text{DRMS} = \sqrt{3}\,\sigma = \text{RMSR}$$

The GPS position dilution of precision (PDOP) is defined in terms of this three-dimensional DRMS (see Appendix D).

5.5.4 Standard Radial Error (SRE)

The variance of the radial error r is

$$\sigma_r^2 = E[(\mathbf{r} - \mu_r)^2] = E[\mathbf{r}^2] - (\mu_r)^2 = \left(3 - \frac{8}{\pi}\right)\sigma^2 \tag{5.14}$$

The standard radial error is the square root of this variance; thus, SRE $= \sigma_r \approx 0.6734\sigma$.

These three error indicators (SRE, RMSR, MRE) are related by

$$\sigma_r^2 = (\text{SRE})^2 = (\text{RMSR})^2 - (\text{MRE})^2 \tag{5.15}$$

Notice that RMSR is not equal to σ or σ_r.

5.5.5 Spherical Error Probable (SEP)

The SEP is defined as the particular radius of the spherical volume V over which the triple integral in Equation (5.7) produces a probability of 50%,

$$p = 0.5 = \int_0^{\text{SEP}} h(r, \sigma)\, dr$$

Figure 5.1 shows the PDF and CDF of the Maxwell distribution. The RMSR, MRE, and SEP points are marked with "*", "+", and "o", respectively. Other points of interest are the 1σ, 2σ, and 3σ probabilities, which are 19.87%, 73.85%, and 97.07%, respectively.

Since the variable r in $H(r)$ is to be interpreted as a normalized quantity ($r = R/\sigma$), the solid CDF curve in Figure 5.1 is valid for all positive values of σ. The variable r in $h(r, \sigma)$ should not be treated as a relative quantity. Thus, the dashed PDF curve in Figure 5.1 is valid only for $\sigma = 1$. In other words, the variable used for the horizontal axis has two interpretations, one for the CDF and another for the PDF.

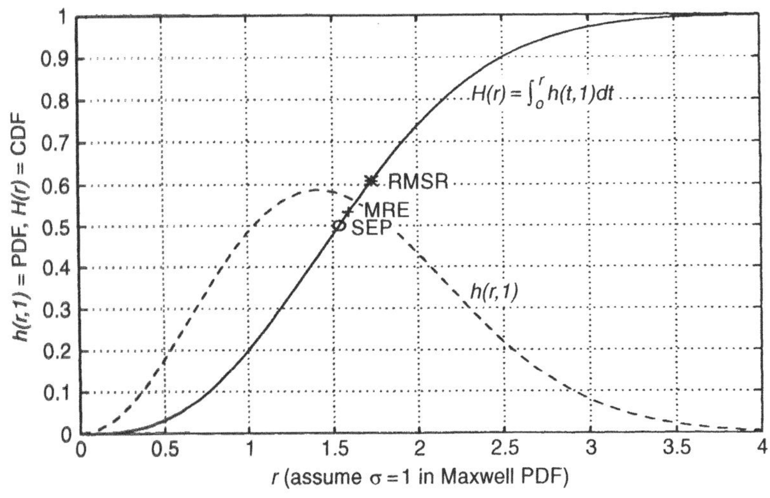

Figure 5.1 PDF and CDF for the Maxwell Distribution.

Figure 5.2 shows the PDF $h(r, \sigma)$ of Equation (5.9) with $\sigma = 1$. The shaded area under the $h(r, 1)$ curve in the [0,1] interval (or $R = 1$) is the probability $p = H(1/1) = -\sqrt{2/\pi}\exp(-1/2) + \text{erf}(1/\sqrt{2}) = 0.1987$, according to Equation (5.10). However, if $\sigma = 2$, then $h(r, 2)$ will be different from that in Figure 5.2 and we have $p = H(1/2) = -\sqrt{2/\pi}\exp(-1/8) + \text{erf}[1/(2\sqrt{2})] = 0.0309$.

The terms 1-SIGMA (1σ) error, 2-SIGMA error, and so on, in three-dimensional error analysis should only be used in the special spherical normal distribution case. In all other cases, it is not even defined. A statement such as "the position accuracy of a system is 60 m (2σ)" could cause the reader to wonder whether the

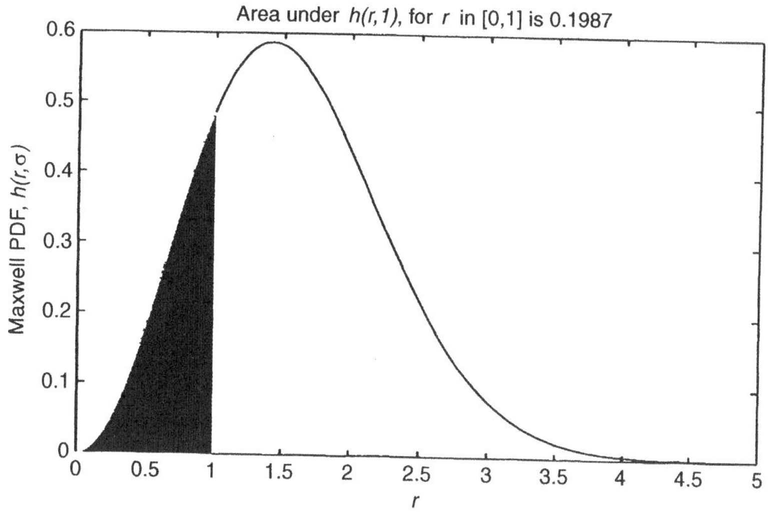

Figure 5.2 $p = \int_0^1 h(r, \sigma)\, dr$ for the Maxwell PDF, $\sigma = 1$.

measure is a spherical error, in which case the numbers describe the 73.85% probability sphere. It can also be interpreted as a one-dimensional sigma along each axis, in which case the 95.45% probability sphere is implied (assuming the distribution to be spherical normal, which actually may not be so). Extra care is needed to avoid confusion.

Table 5.2 contains the R-to-p pairs for the direct problem in this special case. R is expressed in three forms: $R = k \cdot \sigma$, $R = k \cdot \text{MRE}$, and $R = k \cdot \text{SEP}$.

Table 5.3 shows the p-to-R pairs for the inverse problem. In this table, for a given p, R is expressed as multiples of σ, MRE, and SEP; that is, $k = R/\sigma$, $k = R/\text{MRE}$, and $k = R/\text{SEP}$, respectively.

The probabilities for $R = 1 \cdot \text{DRMS}$ and $2 \cdot \text{DRMS}$ are 60.21% and 99.26%, respectively.

For direct problems, we use the M-file **r2p3d.m**, which takes a multiple k of SIGMA, RMSR, MRE, and SEP, and then produces the corresponding probability p for $R = k \cdot \text{SIGMA}$, $R = k \cdot \text{MRE}$, $R = k \cdot \text{RMSR}$, and $R = k \cdot \text{SEP}$, respectively. For example, entering r2p3d(2) in MATLAB results in

```
For R = 2*SIGMA
  p = 0.7385
For R = 2*MRE
  p = 0.9829
For R = 2*RMSR
  p = 0.9926
For R = 2*SEP
  p = 0.9763
```

TABLE 5.2 Three-Dimensional Direct Problem, from R to p

k	$p = H(R/\sigma)$ $R = k \cdot \sigma$	$p = H(R/\sigma)$ $R = k \cdot \text{MRE}$	$p = H(R/\sigma)$ $R = k \cdot \text{SEP}$
0.0	0.0000	0.0000	0.0000
0.1	0.0003	0.0011	0.0010
0.2	0.0021	0.0084	0.0075
0.3	0.0070	0.0273	0.0245
0.4	0.0162	0.0613	0.0554
0.5	0.0309	0.1120	0.1016
0.6	0.0516	0.1786	0.1629
0.7	0.0789	0.2584	0.2372
0.8	0.1128	0.3473	0.3210
0.9	0.1529	0.4405	0.4101
1.0	0.1987	0.5331	0.5000
1.1	0.2494	0.6207	0.5867
1.2	0.3038	0.7003	0.6670
1.3	0.3608	0.7695	0.7384
1.4	0.4192	0.8275	0.7996
1.5	0.4778	0.8745	0.8504
1.6	0.5355	0.9111	0.8911
1.7	0.5911	0.9387	0.9227
1.8	0.6439	0.9589	0.9466
1.9	0.6932	0.9732	0.9639
2.0	0.7385	0.9829	0.9763
2.1	0.7795	0.9895	0.9848
2.2	0.8161	0.9937	0.9905
2.3	0.8482	0.9963	0.9942
2.4	0.8761	0.9979	0.9965
2.5	0.8999	0.9988	0.9980
2.6	0.9200	0.9994	0.9989
2.7	0.9368	0.9997	0.9994
2.8	0.9506	0.9998	0.9997
2.9	0.9617	0.9999	0.9998
3.0	0.9707	1.0000	0.9999
3.1	0.9778	1.0000	1.0000
3.2	0.9834	1.0000	1.0000
3.3	0.9877	1.0000	1.0000
3.4	0.9909	1.0000	1.0000
3.5	0.9934	1.0000	1.0000
3.6	0.9953	1.0000	1.0000
3.7	0.9966	1.0000	1.0000
3.8	0.9976	1.0000	1.0000
3.9	0.9984	1.0000	1.0000
4.0	0.9989	1.0000	1.0000

TABLE 5.3 Three-Dimensional Inverse Problem, from p to R

p	$k = R/\sigma$	$k = R/\text{MRE}$	$k = R/\text{SEP}$
0.000	0.0000	0.0000	0.0000
0.025	0.4645	0.2911	0.3020
0.050	0.5931	0.3717	0.3856
0.075	0.6870	0.4305	0.4466
0.100	0.7644	0.4790	0.4970
0.125	0.8321	0.5214	0.5409
0.150	0.8932	0.5597	0.5807
0.175	0.9496	0.5951	0.6174
0.200	1.0026	0.6283	0.6518
0.225	1.0529	0.6598	0.6845
0.250	1.1012	0.6900	0.7159
0.275	1.1478	0.7193	0.7462
0.300	1.1932	0.7477	0.7757
0.325	1.2377	0.7756	0.8046
0.350	1.2812	0.8029	0.8330
0.375	1.3244	0.8299	0.8610
0.400	1.3672	0.8568	0.8888
0.425	1.4098	0.8835	0.9165
0.450	1.4524	0.9102	0.9442
0.475	1.4951	0.9369	0.9720
0.500	1.5382	0.9639	1.0000
0.525	1.5817	0.9912	1.0283
0.550	1.6257	1.0188	1.0569
0.575	1.6706	1.0469	1.0861
0.600	1.7164	1.0756	1.1159
0.625	1.7635	1.1051	1.1465
0.650	1.8119	1.1355	1.1780
0.675	1.8621	1.1669	1.2106
0.700	1.9144	1.1997	1.2446
0.725	1.9691	1.2340	1.2802
0.750	2.0269	1.2702	1.3177
0.775	2.0884	1.3087	1.3577
0.800	2.1544	1.3501	1.4007
0.825	2.2264	1.3952	1.4474
0.850	2.3059	1.4450	1.4991
0.875	2.3957	1.5013	1.5575
0.900	2.5003	1.5668	1.6255
0.925	2.6277	1.6466	1.7083
0.950	2.7955	1.7518	1.8174
0.975	3.0575	1.9160	1.9878
0.990	3.3682	2.1107	2.1897

For inverse problems, the M-file **p2r3d.m** takes p as input and generates R expressed in terms of SIGMA, RMSR, MRE, and SEP. This may be considered as a conversion tool between these four types of three-dimensional errors. For example, entering p2r3d(0.5) in MATLAB results in

```
For p = 0.5

R = 1.5382*SIGMA

R = 0.9693*MRE

R = 0.8881*RMSR

R = 1*SEP
```

5.6 CASE 3, $V = V_s$, $\rho_{xy} = \rho_{yz} = \rho_{xz} = 0$, $u, v \in [0, 1]$

When the random variables **x**, **y**, and **z** are uncorrelated and have unequal standard deviations, the general joint PDF becomes

$$f(x, y, z) = \frac{1}{\sqrt{(2\pi)^3}\sigma_x\sigma_y\sigma_z}\exp\left[\frac{-1}{2}\left(\frac{x^2}{\sigma_x^2} + \frac{y^2}{\sigma_y^2} + \frac{z^2}{\sigma_z^2}\right)\right] \tag{5.16}$$

Because the three standard errors σ_x, σ_y, and σ_z are not equal, we assume that $\sigma_x \geq \sigma_y \geq \sigma_z$ and use u and v to represent the ratios σ_y/σ_x and σ_z/σ_x, respectively. Equation (5.16) is changed to

$$f(x, y, z) = \frac{1}{\sqrt{(2\pi)^3}uv\sigma_x^3}\exp\left[\frac{-1}{2\sigma_x^2}\left(x^2 + \frac{y^2}{u^2} + \frac{z^2}{v^2}\right)\right] \tag{5.17}$$

Replacing rectangular coordinates with spherical coordinates in Equation (5.3), we obtain

$$p = \iiint_{V_s} f(x, y, z)\,dx\,dy\,dz$$

$$= \int_0^R \int_0^\pi \int_0^{2\pi} f(r\sin\phi\cos\theta, r\sin\phi\sin\theta, r\cos\phi)r^2\sin\phi\,d\theta\,d\phi\,dr \tag{5.18}$$

which, after the substitution of $\lambda = r/\sigma_x$, $u = \sigma_y/\sigma_x$, and $v = \sigma_z/\sigma_x$, simplifies to

$$p = \frac{1}{\sqrt{(2\pi)^3}uv} \cdot \int_0^{2\pi}\int_0^\pi\int_0^{R/\sigma_x} \exp\left\{\frac{-\lambda^2}{2u^2v^2}[u^2v^2\sin^2\phi\cos^2\theta\right.$$

$$\left. + v^2\sin^2\phi\sin^2\theta + u^2\cos^2\phi]\right\}\lambda^2\sin\phi\,d\lambda\,d\phi\,d\theta \tag{5.19}$$

Letting $w = \cos\phi$, then p can be changed to

$$p = \frac{4}{\sqrt{2\pi^3}uv}\int_0^{R/\sigma_x} \lambda^2 \exp\left(\frac{-\lambda^2}{2}\right)\int_0^1 \exp\left[\frac{(v^2-1)\lambda^2 w^2}{2v^2}\right]$$

$$\times \int_0^{\pi/2} \exp\left[\frac{(u^2-1)\lambda^2(1-w^2)\cos^2\theta}{2u^2}\right] d\theta\, dw\, d\lambda \tag{5.20}$$

Thus, p depends not only on the ratio R/σ_x, but also on the ellipticities $u = \sigma_y/\sigma_x$ and $v = \sigma_z/\sigma_x$, which are derivable or known parameters. No analytic solutions are possible for both the direct and the inverse problems in this case. Numerical integration and root searching techniques must be used. The M-files **tf2.m** and **tf3.m** accomplish these goals in normalized form. The MATLAB commands are $p = tf2(r, u, v)$ and $r = tf3(p, u, v)$, respectively. The corresponding general (non-normalized) forms are $p = edsvud(R, [\sigma_x, \sigma_y, \sigma_z])$ and $R = edsvui(p, [\sigma_x, \sigma_y, \sigma_z])$, respectively. For example,

`p = edsvud(2,[1.5,1.6,1])` produces `p = 0.4573,`

`R = edsvui(0.4573,[1.5,1.6,1])` produces `R = 2.0000.`

It takes longer to run **edsvui.m** than **edsvud.m**, as is usually true for inverse problems.

Figure 5.3 shows how the probability p changes when u and v are varied from 0 to 1 for a $R = 1 \cdot$ DRMS. Recall that as u, v vary, 1 DRMS $= \sqrt{1 + u^2 + v^2}\,\sigma_x$ does not remain constant.

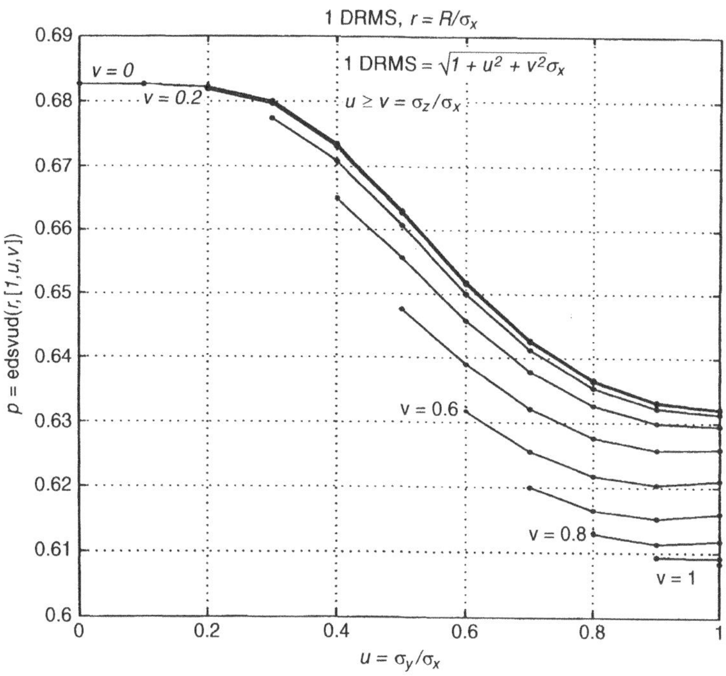

Figure 5.3 Probability Versus u with v as Parameter for $R = 1 \cdot$ DRMS.

Figure 5.4 is similar to Figure 5.3, except that it applies to the case of $R = 2 \cdot \text{DRMS}$.

We notice the resemblance of Figures 5.3 and 5.4 to the two graphs in Figure 4.7 for the two-dimensional case.

The normalized radius $r = R/\sigma_x$, as a function of both ellipticities u, v, is plotted in Figures 5.5 and 5.6 for various fixed probability values $p = 0.5$ and $p = 0.95$, respectively.

We conclude this section with an example to find the radius of a 50% probable error-sphere.

EXAMPLE 5.3

If a certain position error has an ellipsoidal normal distribution with $\sigma_x = 2.5$ m, $\sigma_y = 2$ m, $\sigma_z = 1$ m, and errors in the x, y, z channels are uncorrelated, find the radius of a sphere such that 50% of the time a measured position error will fall inside this sphere.

```
R = edsvui(0.5,[2.5, 2, 1]) = 2.8441
```

Thus, the SEP or $R_{50\%} = 2.8441$ m.

Figure 5.7 shows the volume ($V_s = 4\pi R^3/3$) versus u and v, for $p = 50\%$ and $\sigma_x = 1$, $0 \le u, v \le 1$. The maximum ($V_s = 15.2444$) occurs when $R = 1.5382$, corresponding to $u = v = 1$ or when all σ's equal one.

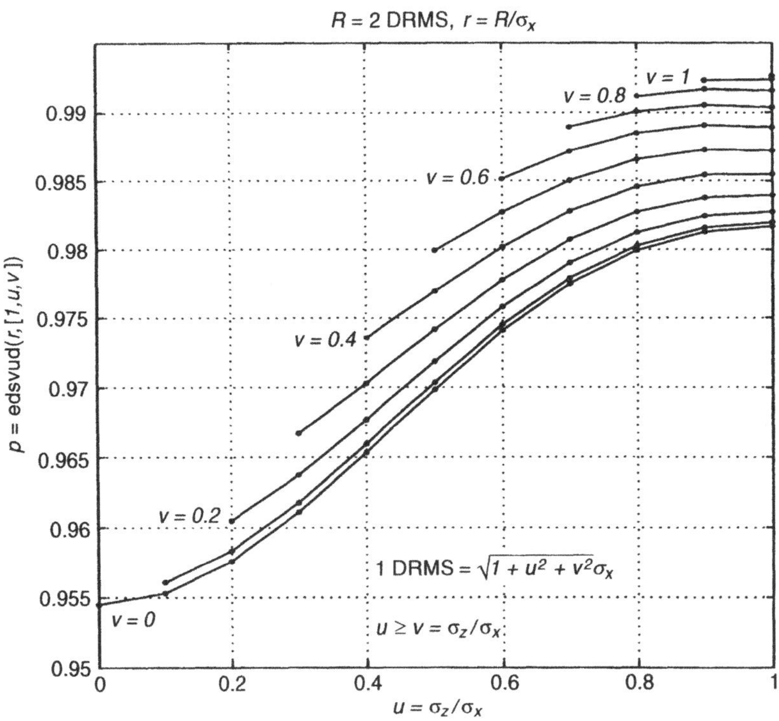

Figure 5.4 Probability Versus u with v as Parameter for $R = 2 \cdot \text{DRMS}$.

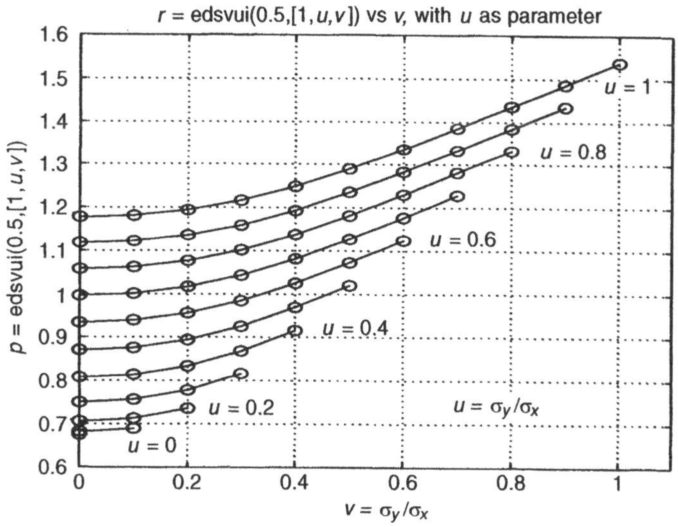

Figure 5.5 Normalized Radius for Various Ellipticities when $p = 0.5$.

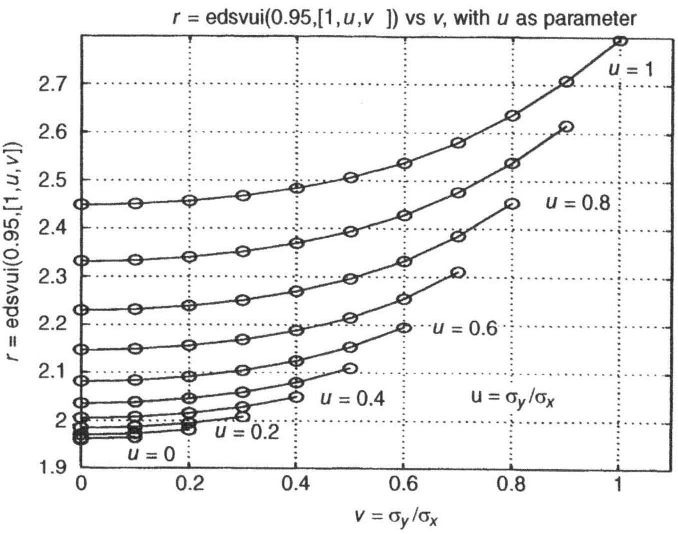

Figure 5.6 Normalized Radius for Various Ellipticities when $p = 0.95$.

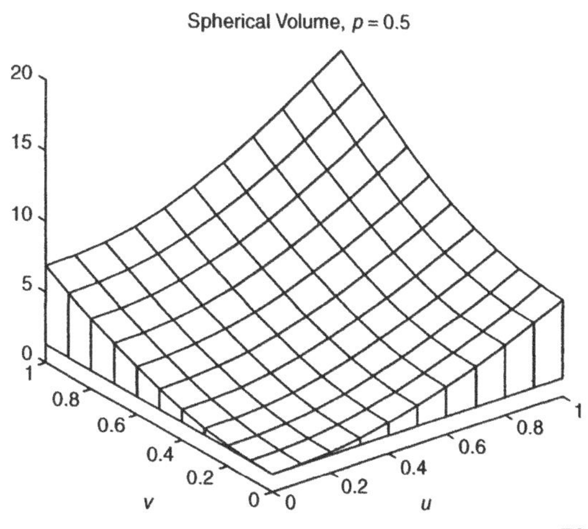

Figure 5.7 Spherical Volume for Various u and v Values when $p = 50\%$.

5.7 CASE 4, $V = V_s$, $\rho_{xy}, \rho_{yz}, \rho_{xz} \in [-1, 1]$, $u, v \in [0, 1]$

Under the assumptions for the present case, Equation (5.3) becomes

$$p = \iiint_{V_s} \frac{1}{\sqrt{(2\pi)^3(1 + 2\rho_{xy}\rho_{yz}\rho_{xz} - \rho_{xy}^2 - \rho_{yz}^2 - \rho_{xz}^2)}\,\sigma_x\sigma_y\sigma_z}$$
$$\times \exp\left\{\frac{-1}{2(1 + 2\rho_{xy}\rho_{yz}\rho_{xz} - \rho_{xy}^2 - \rho_{yz}^2 - \rho_{xz}^2)}\right.$$
$$\times \left[\frac{(1 - \rho_{yz}^2)x^2}{\sigma_x^2} + \frac{(1 - \rho_{xz}^2)y^2}{\sigma_y^2} + \frac{(1 - \rho_{xy}^2)z^2}{\sigma_z^2} - \frac{2(\rho_{xy} - \rho_{xz}\rho_{yz})xy}{\sigma_x\sigma_y}\right.$$
$$\left.\left. - \frac{2(\rho_{yz} - \rho_{xy}\rho_{xz})yz}{\sigma_y\sigma_z} - \frac{2(\rho_{xz} - \rho_{xy}\rho_{yz})xz}{\sigma_x\sigma_z}\right]\right\} dx\,dy\,dz \tag{5.21}$$

with the volume of integration being the interior of a sphere

$$x^2 + y^2 + z^2 \le R^2$$

Let

$$Ax^2 + By^2 + Cz^2 + D_1xy + D_2yz + D_3xz = \frac{1}{(1 + 2\rho_{xy}\rho_{yz}\rho_{xz} - \rho_{xy}^2 - \rho_{yz}^2 - \rho_{xz}^2)}$$
$$\times \left[\frac{(1 - \rho_{yz}^2)x^2}{\sigma_x^2} + \frac{(1 - \rho_{xz}^2)y^2}{\sigma_y^2} + \frac{(1 - \rho_{xy}^2)z^2}{\sigma_z^2} - \frac{2(\rho_{xy} - \rho_{xz}\rho_{yz})xy}{\sigma_x\sigma_y}\right.$$
$$\left. - \frac{2(\rho_{yz} - \rho_{xy}\rho_{xz})yz}{\sigma_y\sigma_z} - \frac{2(\rho_{xz} - \rho_{xy}\rho_{yz})xz}{\sigma_x\sigma_z}\right] \tag{5.22}$$

The correlation between x, y, z errors can be removed by eliminating the three mixed-product terms (D_1xy, D_2yz, D_3xz) in the preceding expression. This is carried out with eigenrotation of the axes or an appropriate similarity transformation on the quadratic form

$$Ax^2 + By^2 + Cz^2 + D_1xy + D_2yz + D_3xz = [x \;\; y \;\; z]\begin{bmatrix} A & D_1/2 & D_3/2 \\ D_1/2 & B & D_2/2 \\ D_3/2 & D_2/2 & C \end{bmatrix}\begin{bmatrix} x \\ y \\ z \end{bmatrix}$$

to obtain

$$Ax^2 + By^2 + Cz^2 + D_1xy + D_2yz + D_3xz$$
$$= [x \;\; y \;\; z]\begin{bmatrix} q_{11} & q_{21} & q_{31} \\ q_{12} & q_{22} & q_{32} \\ q_{13} & q_{23} & q_{33} \end{bmatrix}\begin{bmatrix} A_1 & 0 & 0 \\ 0 & B_1 & 0 \\ 0 & 0 & C_1 \end{bmatrix}\begin{bmatrix} q_{11} & q_{12} & q_{13} \\ q_{21} & q_{22} & q_{23} \\ q_{31} & q_{32} & q_{33} \end{bmatrix}\begin{bmatrix} x \\ y \\ z \end{bmatrix}$$
$$= [\xi \;\; \eta \;\; \zeta]\begin{bmatrix} A_1 & 0 & 0 \\ 0 & B_1 & 0 \\ 0 & 0 & C_1 \end{bmatrix}\begin{bmatrix} \xi \\ \eta \\ \zeta \end{bmatrix} = A_1\xi^2 + B_1\eta^2 + C_1\zeta^2 \tag{5.23}$$

Note that A_1, B_1, C_1 are the eigenvalues of the matrix

$$\begin{bmatrix} A & D_1/2 & D_3/2 \\ D_1/2 & B & D_2/2 \\ D_3/2 & D_2/2 & C \end{bmatrix}$$

and the three corresponding eigenvectors

$$\begin{bmatrix} q_{11} \\ q_{21} \\ q_{31} \end{bmatrix}, \quad \begin{bmatrix} q_{12} \\ q_{22} \\ q_{32} \end{bmatrix}, \quad \text{and} \quad \begin{bmatrix} q_{13} \\ q_{23} \\ q_{33} \end{bmatrix}$$

are used in the transformation from (x, y, z) to (ξ, η, ζ)

$$\begin{bmatrix} \xi \\ \eta \\ \zeta \end{bmatrix} = \begin{bmatrix} q_{11} & q_{12} & q_{13} \\ q_{21} & q_{22} & q_{23} \\ q_{31} & q_{32} & q_{33} \end{bmatrix} \begin{bmatrix} x \\ y \\ z \end{bmatrix}$$

The spherical surface of the integration volume remains a sphere in the new (ξ, η, ζ) coordinate system,

$$\xi^2 + \eta^2 + \zeta^2 = R^2$$

Note that the problem is now expressed in (ξ, η, ζ) coordinates and shows no dependency on the ρ's; it can be solved by the method used in case 3 as

$$p = \iint_{A'_c} \frac{1}{\sqrt{(2\pi)^3}\sigma_\xi \sigma_\eta \sigma_\zeta} \exp\left\{\frac{-1}{2}\left[\frac{\xi^2}{\sigma_\xi^2} + \frac{\eta^2}{\sigma_\eta^2} + \frac{\zeta^2}{\sigma_\zeta^2}\right]\right\} d\xi\, d\eta$$

where $\sigma_\xi^2 = 1/A_1$, $\sigma_\eta^2 = 1/B_1$, $\sigma_\zeta^2 = 1/C_1$, and the volume of integration A'_c is

$$\xi^2 + \eta^2 + \zeta^2 \leq R^2$$

The direct problems are solved by **edsvcd.m**, with the MATLAB command $p = edsvcd(R, [\sigma_x, \sigma_y, \sigma_z], [\rho_{xy}, \rho_{yz}, \rho_{xz}])$, whereas the inverse problems are solved by **edsvci.m** via the command

$$R = edsvcd(p, [\sigma_x, \sigma_y, \sigma_z], [\rho_{xy}, \rho_{yz}, \rho_{xz}]).$$

For example,

`p = edsvcd(2,[1.2,1.4,0.5],[0.5,0.7,0.4])` results in

`p = 0.6866,`

`R = edsvci(0.6866,[1.2,1.4,0.5],[0.5,0.7,0.4])`
results in

`R = 2.0000.`

Notice that when $\rho_{xy} = \rho_{yz} = \rho_{xz} = 0$, case 4 is specialized to case 3; and when both $\rho_{xy} = \rho_{yz} = \rho_{xz} = 0$, and $u = v = 1$, case 4 is reduced to case 2.

5.8 CASE 5, $V = V_e$, $\rho_{xy} = \rho_{yz} = \rho_{xz} = 0$, $u, v \in [0, 1]$

In this case, the probability in Equation (5.3) becomes

$$p = \iiint\limits_{V_e} \frac{1}{2\pi\sigma_x\sigma_y\sigma_z} \exp\left\{\frac{-1}{2}\left[\frac{x^2}{\sigma_x^2} + \frac{y^2}{\sigma_y^2} + \frac{z^2}{\sigma_z^2}\right]\right\} dx\, dy \tag{5.24}$$

where the volume of interest V_e is an ellipsoid

$$\frac{x^2}{a^2} + \frac{y^2}{b^2} + \frac{z^2}{c^2} \leq 1$$

We solve this problem by using a set of scaled variables

$$x_1 = x/a, \quad x_2 = y/b, \quad \text{and } x_3 = z/c$$

so that Equation (5.24) becomes

$$p = \iiint\limits_{V_s'} \frac{1}{2\pi\sigma_1\sigma_2\sigma_3} \exp\left\{\frac{-1}{2}\left[\frac{x_1^2}{\sigma_1^2} + \frac{x_2^2}{\sigma_2^2} + \frac{x_3^2}{\sigma_3^2}\right]\right\} dx_1\, dx_2\, dx_3 \tag{5.25}$$

where $\sigma_1 = \sigma_x/a$, $\sigma_2 = \sigma_y/b$, $\sigma_3 = \sigma_z/c$, and the volume of integration V_s' is a unit sphere in the new coordinates expressed as

$$x_1^2 + x_2^2 + x_3^2 \leq 1$$

Once again we have converted the direct problem of Equation (5.24) to one that is solvable by the method used in case 3.

The M-files **edevud.m** and **edevui.m** have been developed to solve the direct and inverse problems. The file edevud.m includes the needed scaling before calling edsvud.m. The MATLAB commands are

$p = \mathit{edevud}([a, b, c], [\sigma_x, \sigma_y, \sigma_z])$ and $[a, b, c] = \mathit{edevui}(p, k_1, k_2, [\sigma_x, \sigma_y, \sigma_z])$.

Recall that k_1, k_2 are defined as $k_1 = a/b$, $k_2 = b/c$.

For example,

`p = edevud([1.5,1.5,1],[2,2,1.4])`
results in `p = 0.0910,` and

`[a,b,c] = edevui(0.0910,1,1.5,[2,2,1.4])` results in

`[a,b,c] = [1.5000,1.5000,1.0000]`

Notice that when $a = b = c$, case 5 is simplified to case 3; when $a = b = c$ and $u = v = 1$, case 5 is reduced to case 2.

The following example shows how to find a 50%-error ellipsoid.

EXAMPLE 5.4

If a certain position error has an ellipsoidal normal distribution with $\sigma_x = 2.5$ m, $\sigma_y = 2$ m, $\sigma_z = 1$ m, and the errors in the x, y, z channels are uncorrelated, find the semi-axes (a, b, c) of an ellipsoid such that 50% of the time a sample position error will fall within this ellipsoid. Assume that $k_1 = (a/b) = 1.25$, $k_2 = (b/c) = 2$.

The answer can be obtained via the command

```
[a,b,c] = edevui(0.5,1.25,2, [2.5, 2, 1])
        = [3.8454, 3.0763, 1.5382]
```

Thus, the 50%-error ellipsoid has a semi-axis of 3.8454 m in the x direction, 3.0763 m in the y direction, and 1.5382 m in the z direction.

For the same probability, the volume of the error sphere is larger than the volume of the error ellipsoid, as seen from Example 5.3, $V_s = 4\pi R^3/3 = 79.4170$, and from Example 5.4, $V_e = 4\pi abc/3 = 39.9549$.

Figure 5.8 shows the ratio of the volume of the 50%-error-ellipsoid to that of the 50%-error-sphere for various values of u and v. The graph shows that $V_e/V_s \leq 1$ and, as the ellipticities u and v approach 1, the ellipsoidal volume approaches, and eventually coincides with, the spherical volume.

5.8.1 Limiting Situation: $a/\sigma_x = b/\sigma_y = c/\sigma_z$

In general, when $a/\sigma_x = b/\sigma_y = c/\sigma_z$, Equation (5.24) can be greatly simplified to

$$p = -\sqrt{\frac{2}{\pi}}k\exp(-k^2/2) + \text{erf}(k/\sqrt{2}), \quad \text{with } k = a/\sigma_x = b/\sigma_y = c/\sigma_z$$

This equation is similar to Equation (5.9). Thus, an ellipsoidal normal PDF over an ellipsoidal volume with $a/\sigma_x = b/\sigma_y = c/\sigma_z$ can be treated just as the situation of a spherical normal PDF over a sphere.

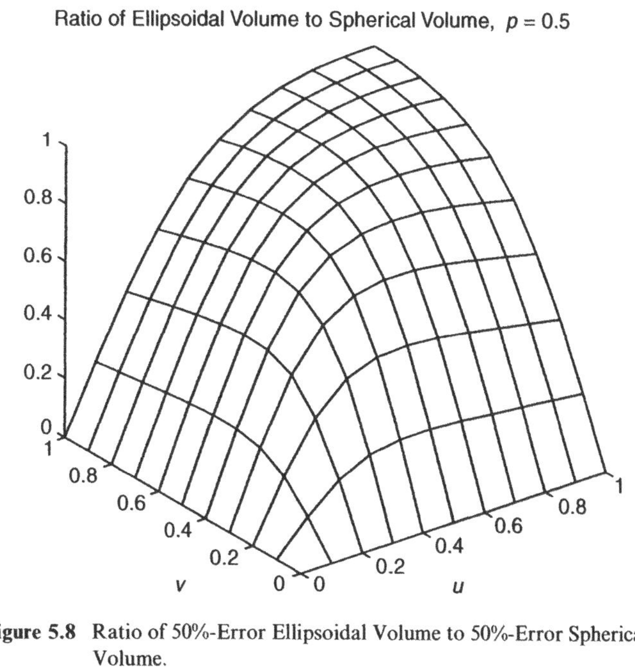

Figure 5.8 Ratio of 50%-Error Ellipsoidal Volume to 50%-Error Spherical Volume.

5.9 CASE 6, $A = A_e$, $\rho_{xy}, \rho_{yz}, \rho_{xz} \in [-1, 1]$, $u, v \in [0, 1]$

The only difference between the present case and case 4 is that the integration area is now within an ellipsoid

$$\frac{x^2}{a^2} + \frac{y^2}{b^2} + \frac{z^2}{c^2} \leq 1$$

instead of a sphere $x^2 + y^2 + z^2 \leq R^2$.

Therefore, if we rescale the variables x, y, and z by

$$x_1 = x/a$$
$$x_2 = y/b$$
$$x_3 = z/c$$

then Equation (5.3) becomes

$$p = \iiint_{V_s'} \frac{1}{\sqrt{(2\pi)^3(1 + 2\rho_{xy}\rho_{yz}\rho_{xz} - \rho_{xy}^2 - \rho_{yz}^2 - \rho_{xz}^2)}\,\sigma_1\sigma_2\sigma_3}$$
$$\times \exp\left\{\frac{-1}{2(1 + 2\rho_{xy}\rho_{yz}\rho_{xz} - \rho_{xy}^2 - \rho_{yz}^2 - \rho_{xz}^2)}\right.$$
$$\times \left[\frac{(1-\rho_{yz}^2)x_1^2}{\sigma_1^2} + \frac{(1-\rho_{xz}^2)x_2^2}{\sigma_2^2} + \frac{(1-\rho_{xy}^2)x_3^2}{\sigma_3^2} - \frac{2(\rho_{xy} - \rho_{xz}\rho_{yz})x_1x_2}{\sigma_1\sigma_2}\right.$$
$$\left.\left. - \frac{2(\rho_{yz} - \rho_{xy}\rho_{xz})x_2x_3}{\sigma_2\sigma_3} - \frac{2(\rho_{xz} - \rho_{xy}\rho_{yz})x_1x_3}{\sigma_1\sigma_3}\right]\right\} dx_1\, dx_2\, dx_3 \tag{5.26}$$

where $\sigma_1 = \sigma_x/a$, $\sigma_2 = \sigma_y/b$, $\sigma_3 = \sigma_z/c$, and the volume of integration V_s' is a sphere in the new (x_1, x_2, x_3) coordinates expressed as

$$x_1^2 + x_2^2 + x_3^2 \leq 1$$

We can see that this equation is almost identical to Equation (5.21), except that radius R there is replaced by 1 here.

The M-files **edevcd.m** and **edevci.m** have been developed to solve the direct and inverse problems, respectively. The MATLAB commands are

$$p = edevcd([a, b, c], [\sigma_x, \sigma_y, \sigma_z], [\rho_{xy}, \rho_{yz}, \rho_{xz}]),$$

and

$$[a, b, c] = edevci(p, k_1, k_2, [\sigma_x, \sigma_y, \sigma_z], [\rho_{xy}, \rho_{yz}, \rho_{xz}]),$$

with $k_1 = a/b$, $k_2 = b/c$. For example,

`p = edevcd([1.5,1.5,1],[2,2,1.4],[0.5,0.5,0])`
results in `p = 0.1006`,

`[a,b,c] = edevci(0.1006,1,1.5,[2,2,1.4],[0.5,0.5,0])`
results in `[a,b,c] = [1.5000,1.5000,1.0000]`.

Notice that when $\rho_{xy} = \rho_{yz} = \rho_{xz} = 0$, case 6 is specialized to case 5; when $a = b$, case 6 is simplified to case 4; when $\rho_{xy} = \rho_{yz} = \rho_{xz} = 0$ and $a = b = c$, case 6 collapses to case 3; when $\rho_{xy} = \rho_{yz} = \rho_{xz} = 0$, $a = b = c$, and $u = v = 1$, case 6 is reduced to case 2. This is the most general case among cases 2 through 6.

5.10 SUMMARY OF M-FILES

The M-files used or generated in this chapter are summarized in Table 5.4. Recall that radius R and the normalized radius r are related by

$$r = R/\sigma$$

The M-file **edsv.m** calls three files (**nf2.m, cf2.m, gf2.m**) developed in Chapter 4 and two files (sf2.m, tf2.m) from this chapter. It is in turn invoked in many other programs (edsvui.m, edsvud.m, etc.).

TABLE 5.4 M-files Used in Chapter 5

File Name	MATLAB Command
nf2.m	$p = nf2(r)$
cf2.m	$p = cf2(r)$
gf2.m	$p = gf2(r, \mu)$
sf1.m	$y = sf1(r)$
sf2.m	$p = sf2(r)$
sf3.m	$r = sf3(p)$
sf3a.m	$y = sf3a(r)$
sf4.m	$r = sf4(p)$
newton.m	$x = newton(x_0, 'fun', 'dfun', tol)$
tf2.m	$p = tf2(r, u, v)$
tf3.m	$r = tf3(p, u, v)$
r2p3d.m	$r2p3d(k)$
p2r3d.m	$p2r3d(p)$
edbvud.m	$p = edbvud([a, b, c], [\sigma_x, \sigma_y, \sigma_z])$
edbvui.m	$[a, b, c] = edbvui(p, k_1, k_2, [\sigma_x, \sigma_y, \sigma_z])$
edsv.m	$p = edsv(r, u, v)$
sdsvud.m	$p = sdsvud(R, \sigma_x)$
sdsvui.m	$R = sdsvui(p, \sigma_x)$
edsvud.m	$p = edsvud(R, [\sigma_x, \sigma_y, \sigma_z])$
edsvui.m	$R = edsvui(p, [\sigma_x, \sigma_y, \sigma_z])$
edsvcd.m	$p = edsvcd(R, [\sigma_x, \sigma_y, \sigma_z], [\rho_{xy}, \rho_{yz}, \rho_{xz}])$
edsvci.m	$R = edsvci(p, [\sigma_x, \sigma_y, \sigma_z], [\rho_{xy}, \rho_{yz}, \rho_{xz}])$
edevud.m	$p = edevud([a, b, c], [\sigma_x, \sigma_y, \sigma_z])$
edevui.m	$[a, b, c] = edevui(p, k_1, k_2, [\sigma_x, \sigma_y, \sigma_z])$
edevcd.m	$p = edevcd([a, b, c], [\sigma_x, \sigma_y, \sigma_z], [\rho_{xy}, \rho_{yz}, \rho_{xz}])$
edevci.m	$[a, b, c] = edevci(p, k_1, k_2, [\sigma_x, \sigma_y, \sigma_z], [\rho_{xy}, \rho_{yz}, \rho_{xz}])$

The flowchart presented in Figure 5.9 describes the relationships between the major M-files developed in this chapter.

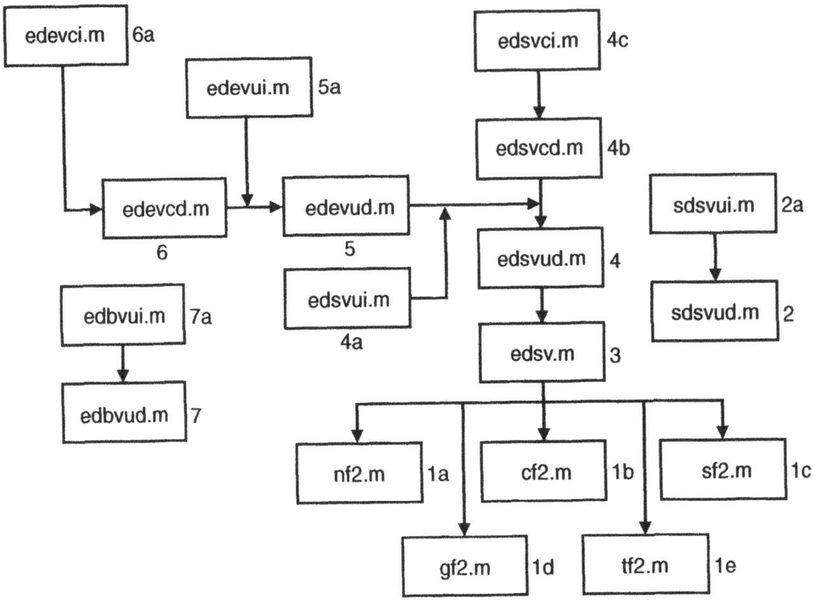

Figure 5.9 Relationships Between Major M-files.

6

Maximum Likelihood Estimation of Radial Error PDF

The method discussed in this chapter for determining the distribution of radial errors differs from other methods (treated in Chapters 4 and 5) in that

it makes no assumption concerning the mean and standard deviation of the normal distribution function of the individual channels or of the correlation existing between them.

The method introduced here is applicable to both two- and three-dimensional cases. The computations required to estimate the radial error PDF are rather involved, but the concepts are simple and straightforward. The original sources of reference are [**11**, **13**].

The basic assumptions are given in the following section along with the general approach. The details of the intermediate steps are presented in the remaining sections.

6.1 BASIC ASSUMPTIONS AND GENERAL APPROACH

Even though no assumptions are made on μ, σ for the normal PDF of the individual channels, an assumption is made regarding the PDF of the radial error.

Assume that the probability density for $(\mathbf{r}^2/a^2)$ can be approximated by the χ_n^2 PDF (chi-square distribution with n degrees of freedom), where "$\mathbf{r}$" is the radial error and "a" is a normalizing factor. Thus, the probability density for $\mathbf{t} = \mathbf{r}^2/a^2$ is given by

$$f(t; n) = \frac{1}{2^{n/2}\Gamma(n/2)} \exp\left[\frac{-t}{2}\right] t^{(n-2)/2}, \quad \text{for } t > 0$$

The general approach to determine the distribution of radial errors is:

- From the N measurements $\{r_1, r_2, \ldots, r_N\}$ of the radial error **r**, compute the geometric mean (**GM**), the mean square (**MS**), and the root mean square (**RMS**) via

$$\mathrm{GM} = \sqrt[N]{\prod_{i=1}^{N} r_i} \tag{6.1}$$

$$\mathrm{MS} = \sum_{i=1}^{N} r_i^2/N \tag{6.2}$$

$$\mathrm{RMS} = \sqrt{\sum_{i=1}^{N} r_i^2/N} \tag{6.3}$$

- Use the maximum likelihood method to estimate the two unknown parameters of this distribution (n and a)—that is, the parametric values for which the probability of obtaining the given set of N observations is a maximum. It should be emphasized that there is no relationship between n (the degrees of freedom) and N (the number of observations). The parameter a is also determined in the maximum likelihood sense after n is found.
- Change the variable from **t** to **r** to obtain the PDF of the radial error **r**.

It will be shown that the best estimate of the radius of the $p\%$ circle (or sphere) is a function of only the geometric mean (GM) and the root mean square (RMS) of the observed radial errors.

6.2 MAXIMUM LIKELIHOOD ESTIMATION OF n, a

Since $\mathbf{t} = \mathbf{r}^2/a^2$, the probability density function for **r** is defined via the probability density function $f(t; n)$ of **t**. Hence,

$$g(r; a, n) = f(r^2/a^2; n)\, dt/dr = \frac{2}{2^{n/2}\Gamma(n/2)} \exp\left[\frac{-r^2}{2a^2}\right] \frac{r^{(n-1)}}{a^n} \tag{6.4}$$

This probability density function for **r** is not completely known because the parameters a and n are to be determined. Maximum likelihood estimation will be used to obtain a and n. We formulate the likelihood function L as

$$L(r_1, r_2, \ldots, r_N; a, n) = \prod_{i=1}^{N} g(r_i; a, n) \tag{6.5}$$

Maximum likelihood estimation (MLE) requires that L, or equivalently

$$\ln[L] = \ln\left[\prod_{i=1}^{N} g(r_i; a, n)\right]$$

be maximized with respect to n and a. Expanding the expression above, we have

$$\ln[L] = N\{(1 - (n/2)\ln(2) - \ln[\Gamma(n/2)] - n\ln(a)\} - \sum_{i=1}^{N} r_i^2/(2a^2) + (n-1)\sum_{i=1}^{N}\ln(r_i)$$

which can be simplified, via the definitions of GM and MS, to

$$\frac{\ln[L]}{N} = \{(1 - (n/2)\ln(2) - \ln[\Gamma(n/2)] - n\ln(a)\} - \frac{\text{MS}}{2a^2} + (n-1)\ln(\text{GM})$$

Defining $Q = \ln[L]/N$, taking partial derivatives $\partial Q/\partial a$ and $\partial Q/\partial n$, and setting these partial derivatives to zero, we obtain, from $\partial Q/\partial a = 0$,

$$\frac{-n}{a} + \frac{\text{MS}}{a^3} = 0$$

$$a^2 = \frac{\text{MS}}{n}$$

or

$$a = \frac{\text{RMS}}{\sqrt{n}} \tag{6.6}$$

Also from $\partial Q/\partial n = 0$,

$$\frac{1}{2}\left\{\ln\left[\frac{(\text{GM})^2}{2a^2}\right] - \psi(n/2)\right\} = 0$$

$$\frac{(\text{GM})^2}{2a^2} = \exp[\psi(n/2)]$$

$$\frac{n(\text{GM})^2}{2\text{MS}} = \exp[\psi(n/2)]$$

or

$$\frac{\text{GM}}{\text{RMS}} = \sqrt{\frac{\exp[\psi(n/2)]}{(n/2)}} \tag{6.7}$$

Note that

$$\psi(\alpha) = \frac{d\ln[\Gamma(\alpha)]}{d\alpha}$$

is the **psi function** (or the **digamma function**), and $\Gamma(\alpha)$ is the gamma function

$$\Gamma(\alpha) = \int_0^\infty u^{\alpha-1}\exp(-u)\,du$$

6.3 DEPENDENCY OF GM/RMS ON n

Equation (6.7) represents a complicated functional relationship between n and GM/RMS since it involves the not-so-familiar ψ function. Given a value of GM/RMS, it is not obvious as to which value of n will be produced. Thus, we resort to a graphic method to show the dependency of GM/RMS on n. This requires the evaluation of $\psi(n/2)$ for various real numbers $n > 0$.

We consider the evaluation of $\psi(n/2)$ for (I) $n =$ positive even integer, (II) $n =$ positive odd integer, and (III) $0 < n < 1$ in the following subsections.

6.3.1 $\psi(n/2)$ for $n =$ Positive Even Integer

If $n = 2m$, then $\psi(n/2)$ is

$$\psi(n/2) = \psi(m) = -\gamma + \sum_{k=1}^{m-1} \tfrac{1}{k} \tag{6.8}$$

with

$$\gamma = \lim_{j\to\infty}\left[\sum_{k=1}^{j}(1/k) - \ln(j)\right] = 0.5772156649 = \text{Euler's constant}.$$

6.3.2 $\psi(n/2)$ for $n =$ Positive Odd Integer

When $n = 2m + 1$, $\psi(n/2)$ can be found through

$$\psi(n/2) = -\gamma - 2\ln(2) + 2\sum_{k=1}^{m} \tfrac{1}{2k-1} \tag{6.9}$$

6.3.3 $\psi(n/2)$ for $0 < n/2 < 1$

Under this condition, we use the definition of $\psi(x)$

$$\psi(x) = \int_0^1 \frac{1 - t^{x-1}}{1 - t}\, dt - \gamma \tag{6.10}$$

The M-file **psinh.m** is the MATLAB tool that computes $\psi(n/2)$ for the three cases above.

The recurrence formula

$$\psi(x) = \psi(x+1) - \frac{1}{x} = \int_0^1 \frac{1 - t^{x}}{1 - t}\, dt - \gamma - \frac{1}{x} \tag{6.11}$$

serves two purposes: (I) it allows us to circumvent the singularity at $t = 1$ in Equation (6.10) when $0 < x < 1$, and (II) it allows us to evaluate $\psi(x) = \psi(x-1) + 1/(x-1)$ for $x > 1$. For example, $\psi(1.2) = \psi(0.2) + 1/(0.2)$.

The reflection formula

$$\psi(x) = \psi(1 - x) + \pi/\tan(x\pi) \tag{6.12}$$

allows us to find $\psi(x)$ for negative values of x. For example, when $x = -0.2$ we have $\psi(-0.2) = \psi(1.2) + \pi/\tan(-0.2\pi)$. This is included here for completeness. For our goal to relate GM/RMS to n, however, we need only to evaluate $\psi(x)$ for $x > 0$.

Numerical procedures to compute $\psi(x)$ for all real x values are implemented in the M-file **psin.m**. The MATLAB command is $y = psin(x)$. To obtain n from

GM/RMS, or GM/RMS from n, with MATLAB, use **gmrms2n.m** or **n2gmrms.m**, respectively.

As n varies in [0.1, 100], the functional dependency of GM/RMS on n, as shown in Equation (6.7), is displayed in Figure 6.1 utilizing a semilog scale.

To use this graph, we first compute the ratio GM/RMS from the N measurements $\{r_1, r_2, \ldots, r_N\}$ and then obtain n from Figure 6.1.

Once n is determined from the graphic method, or via numerical interpolation using **gmrms2n.m**, the value of a can be found from $a = \text{RMS}/\sqrt{n}$ of Equation (6.6). With n and a computed, the PDF of r in Equation (6.4) is completely determined.

The direct problem is to find p from

$$p = \int_0^R g(r; a, n)\, dr$$

for known GM, RMS, and R. The numerical details are carried out in **r2pc6.m**. The MATLAB command is $p = r2pc6(\text{GM}, \text{RMS}, R)$.

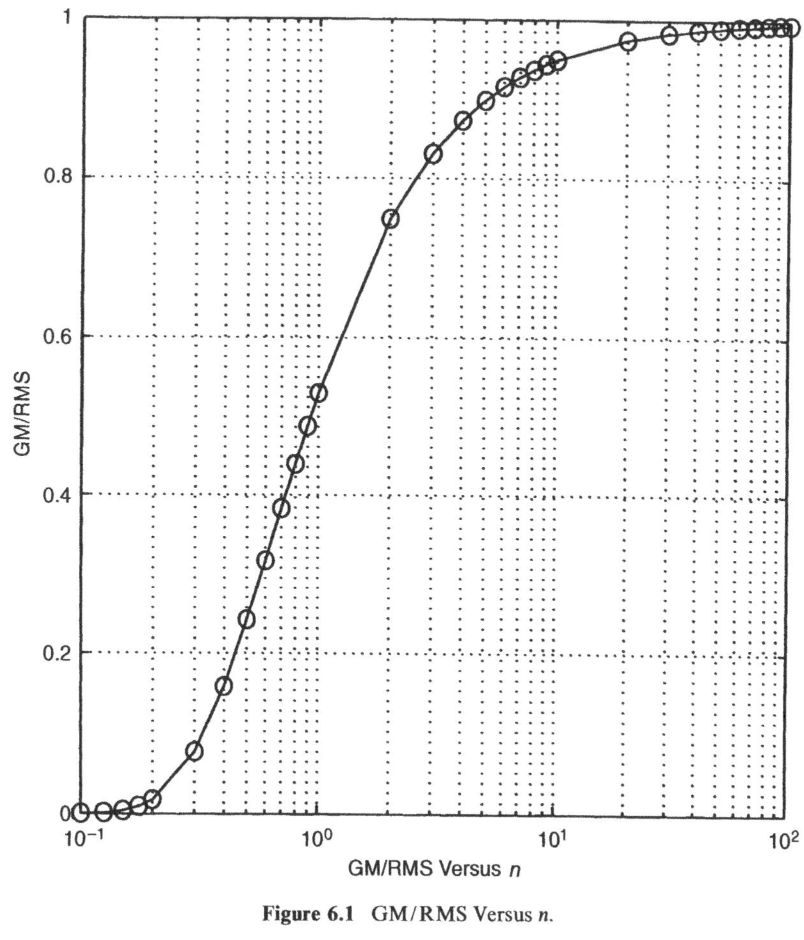

Figure 6.1 GM/RMS Versus n.

The inverse problem is to find $R(p)$ from

$$p = \int_0^{R(p)} g(r; a, n)\, dr$$

for given GM, RMS, and p. The computational steps are carried out in **p2rc6.m**. The MATLAB command is $R = p2rc6(\text{GM}, \text{RMS}, p)$.

6.4 DEPENDENCY OF $R(p)$/RMS ON n

Since $\mathbf{t} = \mathbf{r}^2/a^2$ is assumed to have a χ_n^2 distribution, and $a^2 = \text{MS}/n$, we have

$$t = \frac{r^2}{a^2} = \frac{r^2 n}{\text{MS}}$$

Letting $\tau = \sqrt{\mathbf{t}/n}$, then $\tau = \mathbf{r}/\text{RMS}$ has the χ_n distribution, which is the **chi distribution** with n degrees of freedom:

$$h(\tau; n) = f(n\tau^2; n)\frac{dt}{d\tau} = \frac{2(n/2)^{n/2}\tau^{n-1}\exp(-n\tau^2/2)}{\Gamma(n/2)}$$

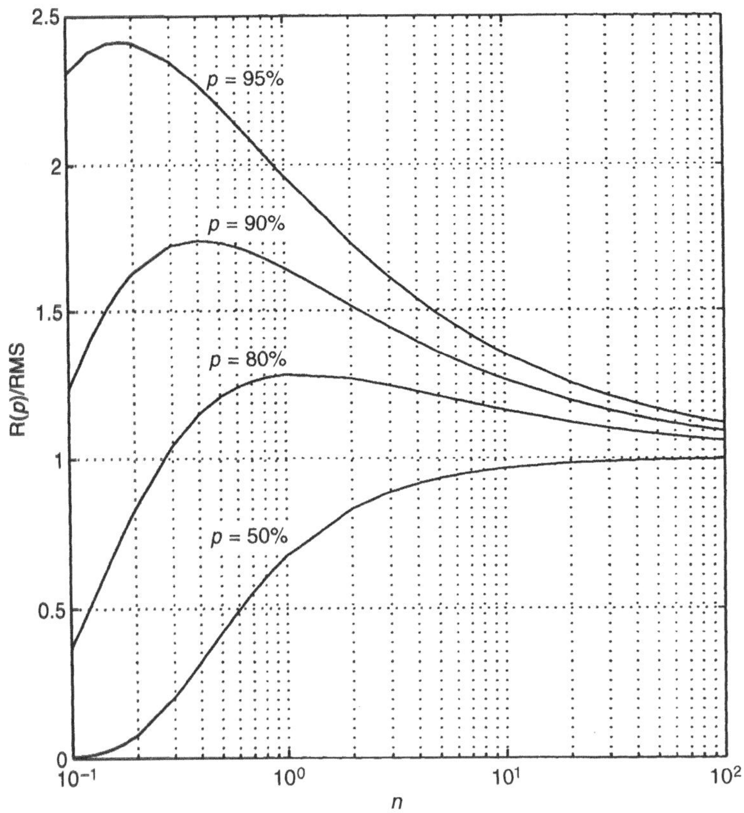

Figure 6.2 $R(p)$/RMS Versus n, for Various p.

Since $t = n\tau^2 = r^2/a^2$, for a given probability p, the following three expressions are equivalent

$$
\begin{aligned}
p &= \int_0^R g(r; a, n)\, dr \\
p &= \int_0^{R/\text{RMS}} h(\tau; n)\, d\tau \\
p &= \int_0^{nR^2/\text{MS}} f(t; n)\, dt
\end{aligned}
\tag{6.13}
$$

Figure 6.2 shows four curves of $R(p)$/RMS versus n for $p = 0.5$, 0.8, 0.9, 0.95. The data for these curves are generated from the M-file **chi1.m** using the last integral in Equation (6.13) with $f(t; n)$ being the χ_n^2 PDF. Notice that we have expanded the range of n, the degrees of freedom for χ^2 PDF, from positive integers to include real numbers in the open interval (0, 1].

6.5 RELATIONSHIP BETWEEN $R(p)$/RMS AND GM/RMS

When combining the information contained in Figures 6.1 and 6.2, we can eliminate n and relate $R(p)$/RMS directly to GM/RMS. The curves showing the relationship between $R(p)$/RMS and GM/RMS for various p values are shown in Figure 6.3.

Here the simplicity of the method is clearly seen. With GM and RMS computable from the N measurements, when p is given (inverse problem), we can obtain $y = R(p)/\text{RMS}$ as the ordinate of the intersection point of the vertical line at $x =$ GM/RMS and the curve corresponding to p. With y and RMS known, we find $R(p) = y \cdot \text{RMS}$. On the other hand, if R is given (direct problem), we locate the point (x, y) on one of the curves. The parameter associated with this particular curve is the probability that the radial error will be found within a circle of radius R. In this case, more curves than shown in Figure 6.3 should be generated.

Of course, both the direct and inverse problems are handled most conveniently by r2pc6.m and p2rc6.m, respectively. We will demonstrate the procedures with three examples.

EXAMPLE 6.1

From the nine observed radial errors shown in Table 6.1 (data are stored in the file c:\mfile\eg6.1), find GM, RMS, and GM/RMS from these observations.

TABLE 6.1 Observed Radial Errors

r_i	2.38	5.18	8.17	13.43	4.43	0.54	6.23	1.08	2.25

The M-file **gmrms.m** is written for this purpose. It is used in MATLAB with a command as

```
[gm, rms, ratio] = gmrms('eg6.1')
```

We obtain GM = 3.3199, RMS = 6.1875, and GM/RMS= 0.5365.

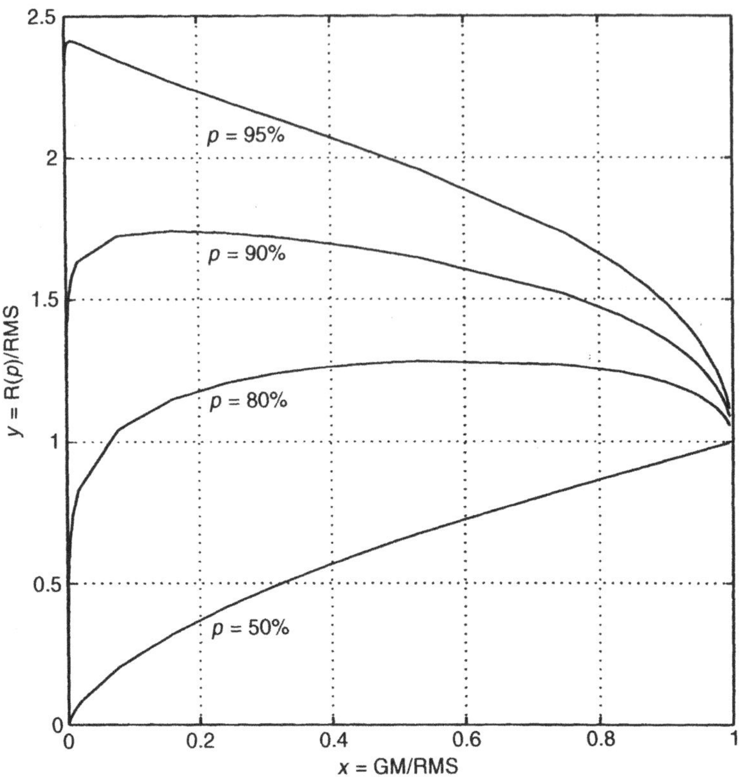

Figure 6.3 $R(p)$/RMS Versus GM/RMS, with p as Parameter.

The next example uses information about errors in the x and y channels.

EXAMPLE 6.2

From the seven pairs of errors in the x and y channels given in Table 6.2 (data are stored in the file c:\mfile\eg6.2), find the $p = 50\%$ point for the radial errors.

TABLE 6.2 Errors in x and y Channels

x_i	y_i
2.38	−13.39
0.18	−24.45
8.17	3.40
3.43	−2.76
4.43	−3.58
0.54	0.49
4.00	−1.03

The geometric mean and root mean square of the radial errors can be solved from the M-file **gmrms.m**. The 50% point of radial error is obtained from p2rc6.m. The MATLAB commands needed are:

```
[gm, rms, ratio] = gmrms('eg6.2');

p = 0.5; R = p2rc6(gm,rms,p);
```

GM, RMS, and GM/RMS are found to be 5.8054, 11.5296, and 0.5035, respectively, and $R = 7.5423$ for $p = 50\%$.

The third example uses information about errors in the x, y, and z channels.

EXAMPLE 6.3

From the eight triplets of errors corresponding to the x, y, and z channels given in Table 6.3 (data are stored in the file c:\mfile\eg6.3), find the probability that the sampled radial errors fall within a circle of radius 10.

TABLE 6.3 Errors in x, y, and z Channels

x_i	y_i	z_i
2.38	−13.39	1.21
0.18	−24.45	−0.35
8.17	3.40	2.44
3.43	−2.76	1.76
4.43	−3.58	0.58
0.54	0.49	−2.14
4.00	−1.03	3.09
−2.24	1.52	2.89

Entering the following three lines in MATLAB results in $p = 61.5\%$ for $R = 10$.

```
[gm, rms, ratio] = gmrms('eg6.3');

R = 10;

p = r2pc6(gm,rms,R);
```

6.6 SUMMARY OF M-FILES

The M-files used or generated in this chapter are summarized in Table 6.4.

The M-file **psi.m** generates the function used as the integrand in Equation (6.11).

The three files **psi1.m**, **chi1.m**, and **chi2.m** compute the data pairs needed to plot the curves in Figures 6.1, 6.2, and 6.3, respectively.

The definite integral for the χ_n^2 PDF, $F(x, n)$, is evaluated using the file **x2cdf.m** which calls **x2df.m** and **x2dg.m**. The command $p = x2cdf(x)$ determines the

TABLE 6.4 M-files Used in Chapter 6

File Name	MATLAB Command
psi.m	$y = psi(x)$
psi1.m	$y = psi1$
psin.m	$y = psin(n)$
psinh.m	$y = psinh(n2)$
x2df.m	$y = x2df(x)$
x2dg.m	$y = x2dg(x)$
x2cdf.m	$p = x2cdf(x)$
x2disc2.m	$y = x2disc2(x)$
chi1.m	*chi1*
chi2.m	*chi2*
gmrms2n.m	$n = gmrms2n(GM, RMS)$
n2gmrms.m	$ratio = n2gmrms(n)$
r2pc6.m	$p = r2pc6(GM, RMS, R)$
p2rc6.m	$R = p2rc6(GM, RMS, p)$
gmrms.m	$[GM, RMS, ratio] = gmrms('filename')$
myfun1.m	$x = myfun1('filename')$

area under the χ_n^2 curve in the interval $[0, x]$. The M-file **x2disc2.m** defines the function $F(x, n) - p$, when the values of x, n, p are supplied.

The M-file **myfun1.m** makes it possible for gmrms.m to call the ASCII data files eg6.1, eg6.2, and eg6.3 (stored in the subdirectory c:\mfile) as input parameters.

7

Position Location Problems

The basic problem in position location is to determine the coordinates of a remote point with respect to a known reference. The remote point may, for example, be a target on which ordnance is to be delivered or a location where troops are to be deployed. Since measurements cannot be made without error, the results of a position determination must be described in terms of the probability of being within a specified distance of the desired point. In general, more than one error is usually involved (for example, x channel and y channel) in the sum total of measurements, and it is necessary to consider the shape of the joint probability density function about the desired point. These functions are usually elliptical normal rather than circular normal. Problems and solutions discussed in this chapter are taken from references [**7, 8**].

An example of a simple position location problem expressed in terms of **error ellipses** is shown in Figure 7.1. Assume a reference baseline is established by the measurement system. The location A of an artillery battery is then measured by the system, and this results in an ellipse within which a given probability is associated. Similarly, the location F of a forward observer has an associated second ellipse within which he may be located to a certain probability. From location F he makes measurements on a target T, which then may be located within still another ellipse. Firing orders in range and azimuth are then calculated based on the above data/information.

The **ballistic dispersion** ellipse is shown in Figure 7.1 as a thick line and is superimposed on the target location ellipse. The goal is to compute the probability of hitting the target.

Since the errors at these three locations (A, F, and T) are expressed by ellipses, and a very elongated ellipse is associated with the weapon, it becomes necessary to develop an analytical procedure that permits ease of analysis when multiple ellipses are involved. We may think of this procedure as a generalization of

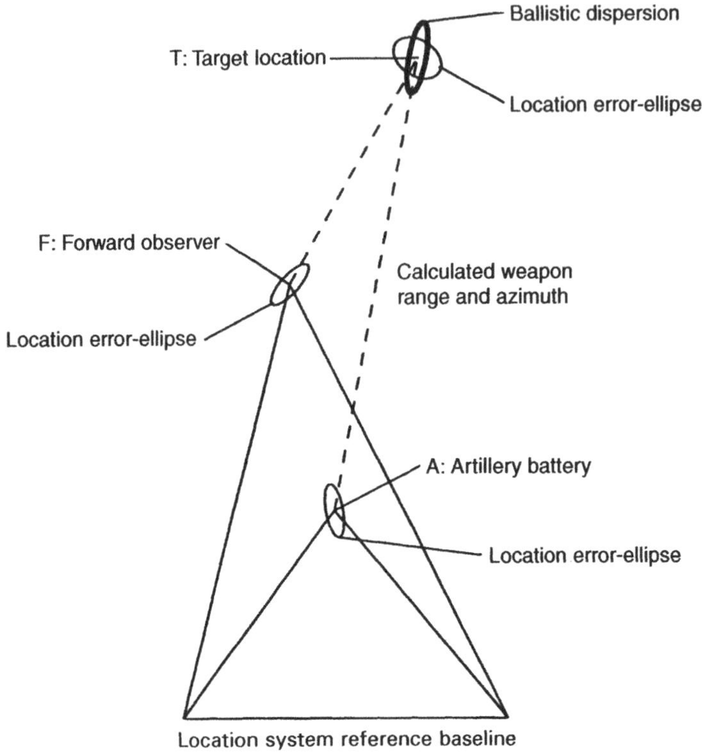

Figure 7.1 Error-Ellipses.

the procedure in Appendix C, which shows how to find the resultant variance after a number of one-dimensional random variables are combined linearly.

To describe an ellipse, we need to know four elements: its center, semi-major and semi-minor axes, and orientation. Hence, any analytical procedure used to obtain the end result must consider and account for a number of error ellipses, including elements of each ellipse, and the angles of these elements with respect to a common reference coordinate system.

We will first study the case of a single error-ellipse. The method of combining several ellipses to obtain the final result will then be described.

7.1 SINGLE ERROR-ELLIPSE ANALYSIS

The determination of a point at the intersection of two lines of position is of primary importance for two-dimensional position location. The lines of position are range measurements (expressed numerically) from two points at the extremities of a baseline with known length.

We will state the main assumptions associated with this analysis and present formulas, graphs, and computer programs to obtain the numerical results. For detailed analysis, the reader is referred to [7].

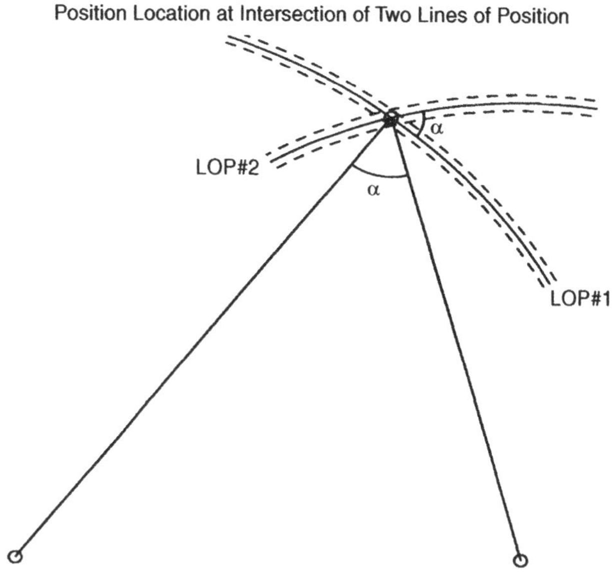

Figure 7.2 Intersection of Two Lines of Position.

Because of measurement error, the actual range obtained from a measurement will not be the true value but will lie within a band as shown by the dashed arcs near the measured lines in Figure 7.2. Thus, one is interested in the probability associated with the actual point (the intersection of the lines of position). It can be shown that the contours of equal probability density about such an intersection point are ellipses centered about the true intersection.

In order to continue with this analysis, a number of assumptions can be made:

- The random error is normally distributed.
- The mean error is zero, or the mean error has been removed.
- The errors associated with the two lines of position (LOP) are independent, and the two standard errors are σ_1 and σ_2.
- The standard deviation is small compared to the actual radius of curvature of the lines of position. Thus, the lines of position can be regarded as straight lines in the neighborhood of the intersection point, as seen in Figure 7.3.

In general, the exact shape of the error figure varies with the magnitude of the standard error associated with each line of position and the angle of intersection between them. We will use σ_1 and σ_2 for the standard errors and α for the **cut angle** (or intersection angle measured counterclockwise from line of position #1 to line of position #2) as shown in the following discussion.

Two separate but interrelated methods can be used to obtain the desired result. The end result is the determination of the probability that the point is located within a circle of stated radius. The basis of this concept is best seen by considering for a moment the special case when the two standard errors are equal ($\sigma_1 = \sigma_2$) and the angle of intersection of the lines of position is a right angle ($\alpha = 90°$). In this case,

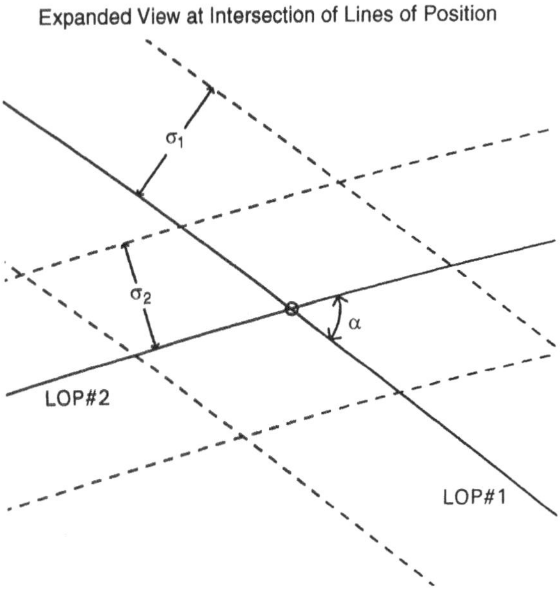

Figure 7.3 Expanded View at Intersection Point.

the error figure becomes a circle and is described by the circular normal distribution; that is, Rayleigh distribution

$$h(t, \sigma) = \frac{t}{\sigma^2} \cdot \exp\left(\frac{-t^2}{2\sigma^2}\right) \tag{7.1}$$

with $t^2 = x^2 + y^2$ and $\sigma = \sigma_1 = \sigma_2$.

Using the normalized variable $r = R/\sigma$, we have the CDF

$$p = H(r) = H(R/\sigma) = 1 - \exp\left(\frac{-r^2}{2}\right) \tag{7.2}$$

In Figure 7.4 both the PDF (for $\sigma = 1$) and the CDF of this random variable $\mathbf{r}$ are plotted. This graph is identical to Figure 4.5.

EXAMPLE 7.1

A measurement system produces circular errors and has a single-axis σ value of 5 m; the probability of actually being located inside a circle with radius $R = 5$ m (implying $r = R/\sigma = 1$) may be read from the $H(r)$ curve of Figure 7.4 to be 0.3934 or 39.34%. To find the CEP, the radius of a circle within which a 50% probability results, the corresponding value for $r = R/\sigma$ is seen from the horizontal axis to be 1.1774. Thus, for this particular example, the CEP would be $r \times \sigma = 1.1774 \times 5 = 5.887$ m. One can also use the M-files **cf2.m** and **cf3.m** to obtain the same results via the commands `p = cf2(1)` and `p = cf3(0.5)`.

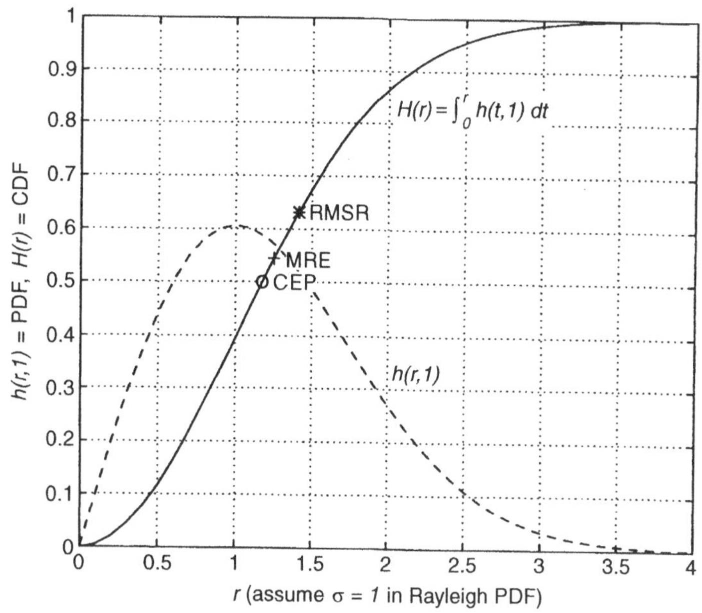

Figure 7.4 PDF and CDF for the Rayleigh Distribution.

7.1.1 Approach 1. Fictitious Sigma and Cut Angle

In this approach, it is assumed that we can find fictitious values of σ such that the two different σ_1 and σ_2, originally given, may be replaced by two new σ's of equal value, indicated as σ^*. Thus, a new and fictitious angle of intersection α^* replaces the old one α. Figure 7.5 illustrates the original and new values for the standard errors and the cut angle.

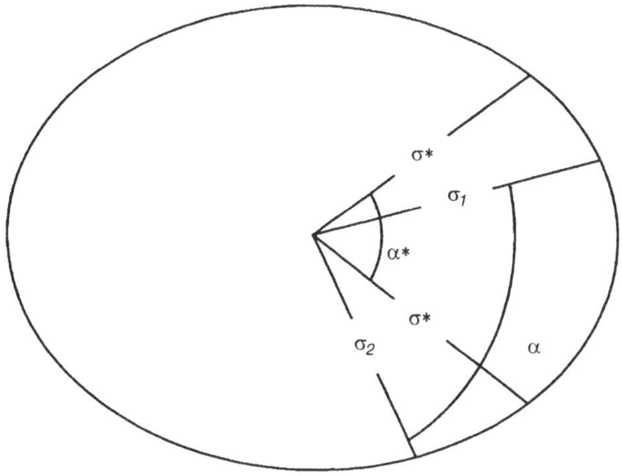

Figure 7.5 Old and New Standard Deviations and Cut Angles.

The values of σ^* and α^* are computed from the two formulas

$$\sigma^* = \frac{\sin(2\beta) \cdot \sqrt{\sigma_1^2 + \sigma_2^2}}{\sqrt{2}} \tag{7.3}$$

$$\alpha^* = \arcsin[\sin(2\beta) \cdot \sin(\alpha)] \tag{7.4}$$

with $\beta = \arctan(\sigma_2/\sigma_1)$. We also define the **sigma factor** as $F_\sigma = \sigma^*/\max(\sigma_1, \sigma_2)$.

These two equations plus sigma factor may be computed using the M-file **fsca.m**. To obtain σ^* and α^* from σ_1, σ_2, and α, simply enter $fsca(\sigma_1, \sigma_2, \alpha)$ in MATLAB. For example, entering *fsca*(2, 3, 30) or *fsca*(3, 2, 30) will produce $\sigma^* = 2.3534$, $\alpha^* = 27.49°$, $F_\sigma = 0.7845$.

If $\sigma_1 \leq \sigma_2$, Figure 7.6 shows the sigma factor $F_\sigma = \sigma^*/\sigma_1$ as a function of the ratio $u = \sigma_2/\sigma_1$.

Figure 7.7 displays the fictitious cut angle α^* versus the original cut angle α, with $u = \sigma_2/\sigma_1$ as a parameter.

Figure 7.8 shows the probability that a measurement lies within a circle of radius $r = R/\max(\sigma_x, \sigma_y)$ under the new joint probability density function

$$f(x, y) = \frac{1}{2\pi\sigma_x\sigma_y} \cdot \exp\left[\frac{-1}{2}\left(\frac{x^2}{\sigma_x^2} + \frac{y^2}{\sigma_y^2}\right)\right] \tag{7.5}$$

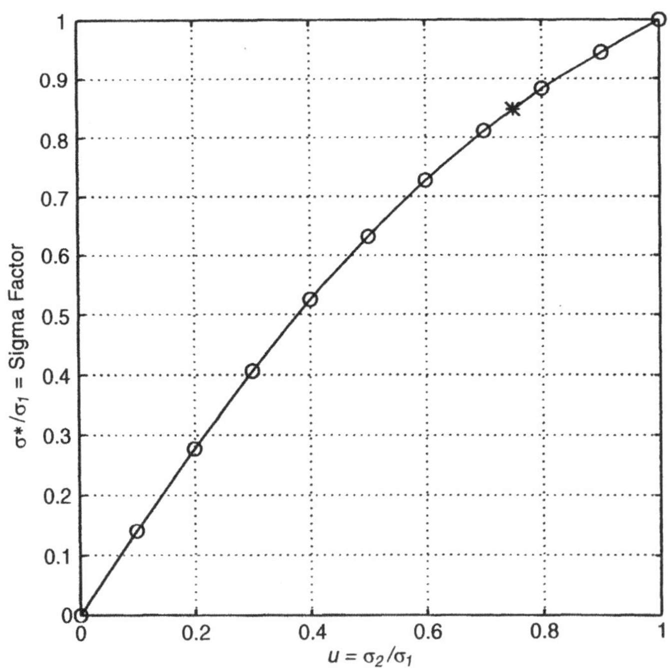

Figure 7.6 Sigma Factor Versus Ellipticity *u*.

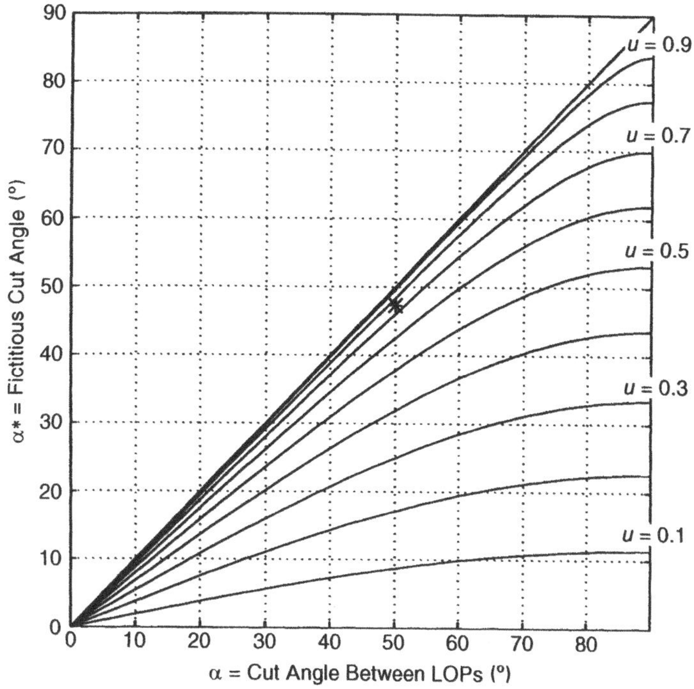

Figure 7.7 Fictitious Cut Angle Versus Original Cut Angle.

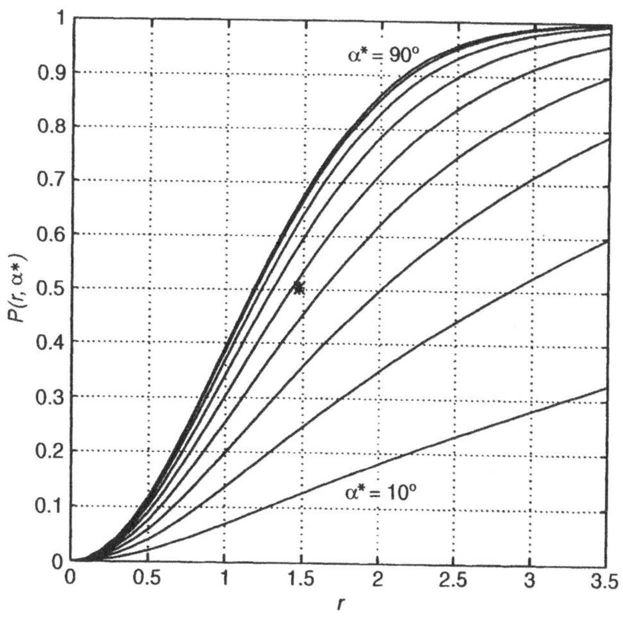

Figure 7.8 Probability Versus r, with Cut Angle as Parameter.

where

$$\sigma_x = \frac{\sigma^*}{\sqrt{2} \cdot \sin(\alpha^*/2)} \quad \text{and} \quad \sigma_y = \frac{\sigma^*}{\sqrt{2} \cdot \cos(\alpha^*/2)}$$

with the fictitious variable σ^* computed from Equation (7.3) and σ^* from Equation (7.4), using the original cut angle and the unequal standard errors σ_1 and σ_2.

The next example shows the procedure for calculation.

EXAMPLE 7.2

Assume that a position location system has provided the following data: $\alpha = 50°$, $\sigma_1 = 20$ m, and $\sigma_2 = 15$ m. What is the probability that the location of the point is within a circle of 25 m radius?

- Start MATLAB.
- Enter $p = seml(25, 15, 20, 50)$ or $p = seml(25, 20, 15, 50)$
  Note that the order of appearance for the two σ's in the input argument list of **seml.m** is reversible.
- $p = 0.5035$ or 50.35% is the answer.

Figures 7.6–7.8 can be used to solve problems similar to that shown in Example 7.2 for those who prefer approximate solutions from graphs rather than precise solutions from MATLAB M-files.

Locations marked with the symbol "*" in Figures 7.6–7.8 correspond to steps 2, 4, and 6, in the following procedure:

1. Compute the ratio $u = \sigma_2/\sigma_1 = 15/20 = 0.75$.
2. Find the sigma factor $F_\sigma = 0.8485$ from Figure 7.6, for $u = 0.75$.
3. Obtain $\sigma^* = F_\sigma \cdot \sigma_1 = 0.8485 \times 20 = 16.97$ m.
4. Find $\alpha^* = 47.34°$ from Figure 7.7, for $u = 0.75$ and $\alpha = 50°$.
5. Find $r = R/\sigma^* = 25/16.97 = 1.473$.
6. Find $p = 0.5035$ from Figure 7.8, for the fictitious cut angle $\alpha^* = 47.34°$ and normalized radius $r = 1.473$.

7.1.2 Approach 2. New Sigmas Along the Major and Minor Axes

A different transformation is used to find new sigmas along the two principal axes of the ellipse from the given sigmas and cut angle (see Figure 7.9). The two new sigmas σ_x, σ_y are computed from

$$\sigma_x^2 = \frac{1}{2\sin^2(\alpha)} \cdot \left[\sigma_1^2 + \sigma_2^2 + \sqrt{\sigma_1^2 + \sigma_2^2 - 4\sin^2(\alpha)\,\sigma_1^2\,\sigma_2^2}\right] \tag{7.6}$$

$$\sigma_y^2 = \frac{1}{2\sin^2(\alpha)} \cdot \left[\sigma_1^2 + \sigma_2^2 - \sqrt{\sigma_1^2 + \sigma_2^2 - 4\sin^2(\alpha)\,\sigma_1^2\,\sigma_2^2}\right] \tag{7.7}$$

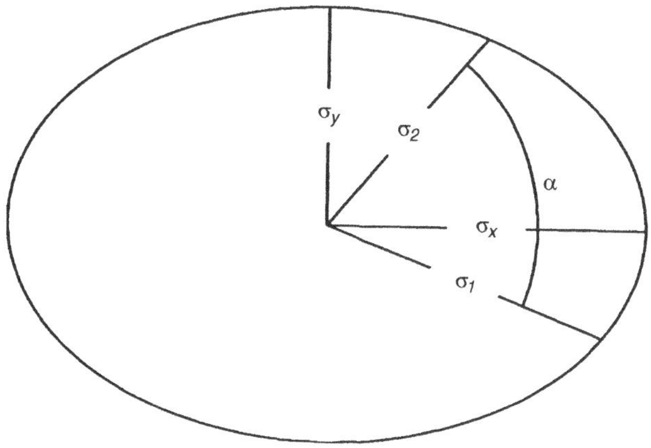

Figure 7.9 Old σ's with Cut Angle and New σ's Along Orthogonal Axes.

We demonstrate this second approach using the same information as given in Example 7.2.

- Start MATLAB.
- Enter $p = sem2(25, 15, 20, 50)$ or $p = sem2(25, 20, 15, 50)$ and obtain $\sigma_x = 29.89$, $\sigma_y = 13.1$, and $r = 0.8364$.
- $p = 0.5035$ or 50.35% is the answer.

Alternately, we can use graphic approximation as follows:

1. Compute $\sigma_x = 29.89$ and $\sigma_y = 13.1$ from Equations (7.6–7.7).
2. Find ellipticity $u = \sigma_y/\sigma_x = 0.4382$.
3. Obtain $r = R/\sigma_x = 0.8364$.
4. Find $p = 0.5035$ from Figure 7.10.

Figure 7.10 shows the probability versus radius curves for various ellipticity values. This graph is used for direct problems.

Figure 7.11 shows the radius of the circle with specified probability as a function of ellipticity. The values used for p are 50%, 60%, 70%, 80%, 90%, and 95%. This graph is used for inverse problems.

The radius r is a normalized quantity (normalized with respect to the maximum of σ_x, σ_y). To use either one of these two figures, we need to compute the ellipticity $u = \min(\sigma_x, \sigma_y)/\max(\sigma_x, \sigma_y)$.

7.2 CONSIDERATION OF GEOMETRICAL FACTORS

Notice that the cut angle plays an important role in each of the two methods described earlier. We will develop curves for constant values of initial error showing

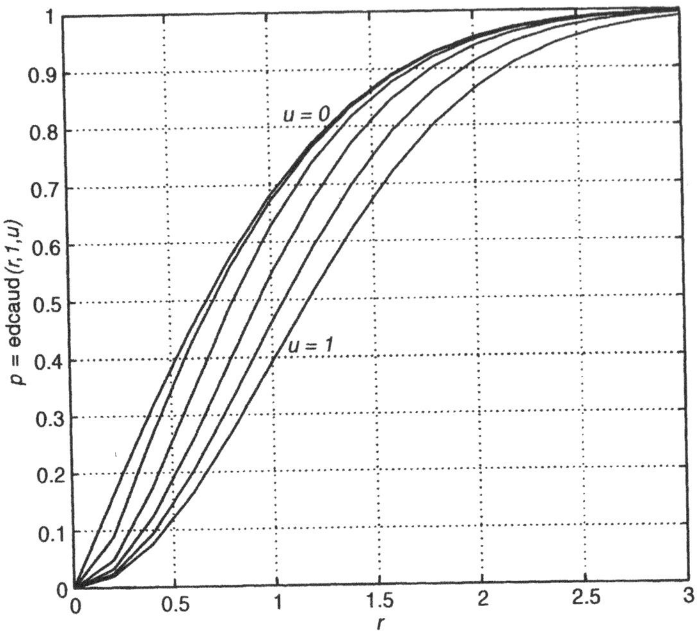

Figure 7.10 Probability Versus r with Ellipticity u as Parameter.

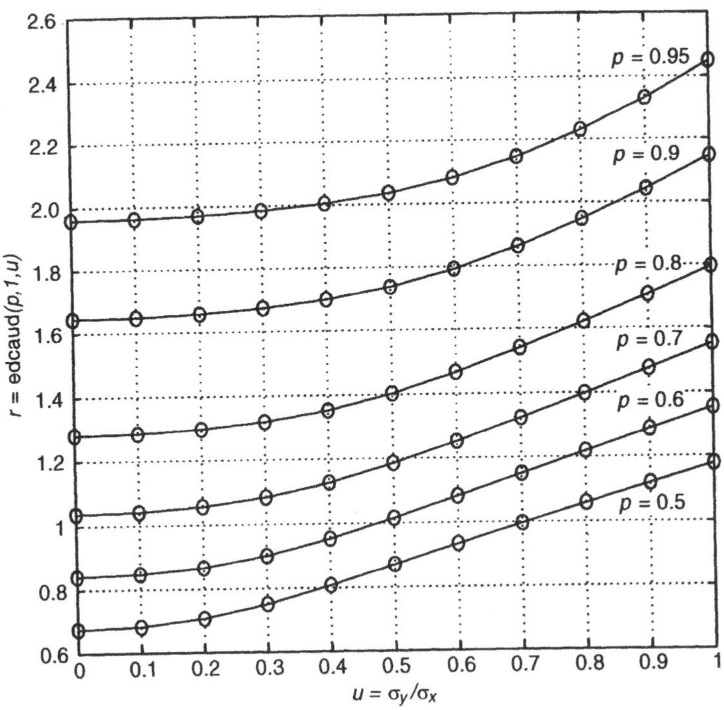

Figure 7.11 Radius of Circle with Specified Probability Versus Ellipticity u.

that the radius of a circle with a fixed value of probability varies as a function of the intersection angle of the lines of position.

To simplify the investigation of geometrical factors, let us assume $\sigma_1 = \sigma_2 = \sigma$. In this case, the formulas for σ_x and σ_y reduce to

$$\sigma_x = \frac{\sigma}{\sqrt{2} \cdot \sin(\alpha/2)} \quad \text{and} \quad \sigma_y = \frac{\sigma}{\sqrt{2} \cdot \cos(\alpha/2)}$$

Hence, the ellipticity is $u = \tan(\alpha/2)$ or $u = \cot(\alpha/2)$, depending on the sizes of σ_x and σ_y.

In any problem involving position, it is well known that the best results are obtained when the lines of position intersect at right angles. We are interested in the impact when the cut angle is not 90°.

TABLE 7.1 Probability Versus Cut Angle for Constant Radius

α (°)	σ_x	σ_y	u	$p[r(50\%, 90°)]$ (%)	$p[r(90\%, 90°)]$ (%)
5	16.2108	0.7078	0.0437	4.50	9.89
10	8.1131	0.7098	0.0875	8.97	19.60
15	5.4174	0.7132	0.1317	13.35	28.96
20	4.0721	0.7180	0.1763	17.61	37.81
25	3.2670	0.7243	0.2217	21.72	46.03
30	2.7321	0.7321	0.2679	25.63	53.54
35	2.3515	0.7414	0.3153	29.33	60.28
40	2.0674	0.7525	0.3640	32.77	66.22
45	1.8478	0.7654	0.4142	35.95	71.38
50	1.6732	0.7802	0.4663	38.83	75.78
55	1.5314	0.7972	0.5206	41.41	79.46
60	1.4142	0.8165	0.5774	43.66	82.50
65	1.3160	0.8384	0.6371	45.59	84.94
70	1.2328	0.8632	0.7002	47.17	86.84
75	1.1615	0.8913	0.7673	48.40	88.26
80	1.1001	0.9231	0.8391	49.29	89.24
85	1.0467	0.9591	0.9163	49.82	89.81
90	1.0000	1.0000	1.0000	50.00	90.00
95	1.0467	0.9591	0.9163	49.82	89.81
100	1.1001	0.9231	0.8391	49.29	89.24
105	1.1615	0.8913	0.7673	48.40	88.26
110	1.2328	0.8632	0.7002	47.17	86.84
115	1.3160	0.8384	0.6371	45.59	84.94
120	1.4142	0.8165	0.5774	43.66	82.50
125	1.5314	0.7972	0.5206	41.41	79.46
130	1.6732	0.7802	0.4663	38.83	75.78
135	1.8478	0.7654	0.4142	35.95	71.38
140	2.0674	0.7525	0.3640	32.77	66.22
145	2.3515	0.7414	0.3153	29.33	60.28
150	2.7321	0.7321	0.2679	25.63	53.54
155	3.2670	0.7243	0.2217	21.72	46.03
160	4.0721	0.7180	0.1763	17.61	37.81
165	5.4174	0.7132	0.1317	13.35	28.96
170	8.1131	0.7098	0.0875	8.97	19.60
175	16.2108	0.7078	0.0437	4.50	9.89

TABLE 7.3 Significant Parameters of Error-Ellipse, $p = 90\%$

α (°)	σ_x	σ_y	u	$r(90\%, \alpha)$	$R(90\%, \alpha)$	$R(90\%, \alpha)/R(90\%, 90°)$
10	8.113	0.710	0.087	1.647	13.364	6.227
15	5.417	0.713	0.132	1.650	8.940	4.166
20	4.072	0.718	0.176	1.654	6.737	3.139
25	3.267	0.724	0.222	1.660	5.424	2.527
30	2.732	0.732	0.268	1.668	4.556	2.123
35	2.351	0.741	0.315	1.677	3.944	1.838
40	2.067	0.752	0.364	1.689	3.492	1.627
45	1.848	0.765	0.414	1.704	3.148	1.467
50	1.673	0.780	0.466	1.722	2.882	1.343
55	1.531	0.797	0.521	1.747	2.675	1.246
60	1.414	0.816	0.577	1.778	2.514	1.171
65	1.316	0.838	0.637	1.816	2.390	1.114
70	1.233	0.863	0.700	1.863	2.296	1.070
75	1.162	0.891	0.767	1.918	2.228	1.038
80	1.100	0.923	0.839	1.983	2.182	1.017
85	1.047	0.959	0.916	2.059	2.155	1.004
90	1.000	1.000	1.000	2.146	2.146	1.000
95	1.047	0.959	0.916	2.059	2.155	1.004
100	1.100	0.923	0.839	1.983	2.182	1.017
105	1.162	0.891	0.767	1.918	2.228	1.038
110	1.233	0.863	0.700	1.863	2.296	1.070
115	1.316	0.838	0.637	1.816	2.390	1.114
120	1.414	0.816	0.577	1.778	2.514	1.171
125	1.531	0.797	0.521	1.747	2.675	1.246
130	1.673	0.780	0.466	1.722	2.882	1.343
135	1.848	0.765	0.414	1.704	3.148	1.467
140	2.067	0.752	0.364	1.689	3.492	1.627
145	2.351	0.741	0.315	1.677	3.944	1.838
150	2.732	0.732	0.268	1.668	4.556	2.123
155	3.267	0.724	0.222	1.660	5.424	2.527
160	4.072	0.718	0.176	1.654	6.737	3.139
165	5.417	0.713	0.132	1.650	8.940	4.166
170	8.113	0.710	0.087	1.647	13.364	6.227

its major and minor axes, the problem is to properly combine more than one ellipse. Determining the proper method needed to combine a number of individual errors in order to obtain the total error at some desired point is discussed later in this chapter.

In the general case, the ellipses will be randomly oriented relative to one another. The exact consideration of this random orientation of axes complicates the analysis but is necessary in order to obtain the correct result.

7.3.1 Mutually Parallel Axes

It will be helpful to investigate the special and unlikely case where the major axis of every ellipse involved is parallel (thus all minor axes are also mutually parallel). The analysis for the more general case will then be considered.

From the problem formulation in Figure 7.1, there are four error-ellipses of interest (with standard deviations for the two axes of each ellipse listed):

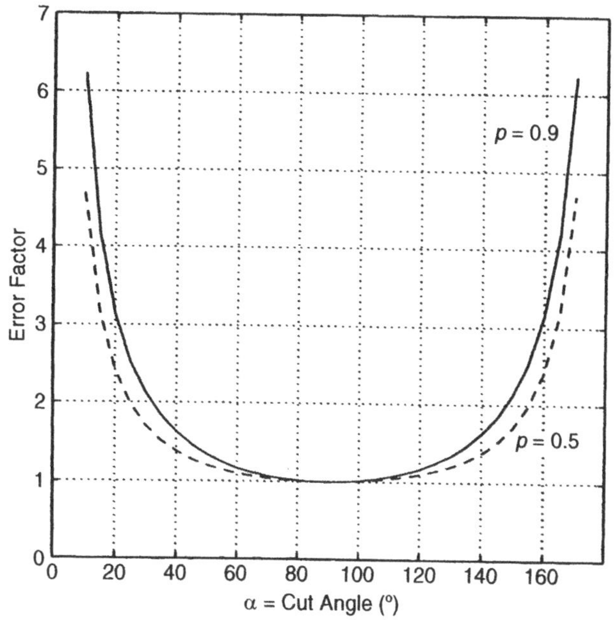

Figure 7.13 Error Factor Versus Cut Angle.

- Weapon dispersion with $\sigma_x = 3$ m, $\sigma_y = 40$ m
- Gun location with $\sigma_x = 10$ m, $\sigma_y = 15$ m
- Forward observer location with $\sigma_x = 15$ m, $\sigma_y = 20$ m
- Target location with respect to forward observer with $\sigma_x = 30$ m, $\sigma_y = 10$ m

We wish to obtain the probability of damage to the target assuming the shell must land within a circle of 20-m radius in order to meet the desired damage level.

For this example, it is assumed that each error-ellipse has its axes mutually parallel and, without loss of generality, aligned north and east. The method of obtaining the total error at the target consists of finding the sum of the variances in the two directions and converting these two sums to the standard deviations of the total error-ellipse at the target.

The computation steps are as follows:

- $\sigma_x = \sqrt{3^2 + 10^2 + 15^2 + 30^2} = \sqrt{1234} = 35.1283$ m
- $\sigma_y = \sqrt{40^2 + 15^2 + 20^2 + 10^2} = \sqrt{2325} = 48.2183$ m
- $u = \min(\sigma_x, \sigma_y)/\max(\sigma_x, \sigma_y) = 35.1283/48.2183 = 0.7285$
- $r = R/\max(\sigma_x, \sigma_y) = 20/48.2183 = 0.4148$
- Use **bf2.m** by entering $p = bf2(r, u)$ in MATLAB to obtain $p = 11.11\%$. One can also use **gf2.m**, first introduced in Chapter 4, to accomplish the same goal.

Each of these steps is implemented in **noval.m** for a system with N error-ellipses; that is, the command $p = noval(R)$, with $R = 20$, will carry out all five steps to determine p. This program requires an ASCII input file **nsd.dat** to store the two standard deviations for each of the N error-ellipses. For the example above, $N = 4$, and the input file nsd.dat consists of

```
 3  40
10  15
15  20
30  10
```

It is interesting to compare the contribution to total error from the dispersion of the weapon to that contributed by location measurement errors of the gun, the forward observer, and the target. Table 7.4 contains the individual contributions from each of these four sources.

Thus, with perfect location of all elements, the gun (with the stated dispersion) has a 37.89% probability of landing a shell within a circle of 20-m radius. When the three location errors are taken into account, however, the probability falls to 11.11% when the gun dispersion is considered.

This method of adding variances along the two orthogonal axes (down range and cross range) is the standard method of preparing an error budget for a weapon system. However, it is not sufficient when one wishes to combine error-ellipses having random orientations of their axes. Since statistical distributions are involved, simple trigonometric resolution from one set of axes to another is not appropriate.

7.3.2 Randomly Oriented Axes

In the general case of N error-ellipses with randomly oriented axes, a more complex procedure for combination is needed. Briefly, a reference set of axes must be chosen, and the orientation of each error-ellipse relative to these axes must be determined. Subsequently, the standard deviations along these axes must be computed, which is a procedure that involves cross-product terms. The formulas for this calculation are given below without derivation **[8]**.

Figure 7.14 illustrates the situation. The three smaller ellipses are the inputs to the problem, and the large ellipse represents the final combination of the three smaller ones.

In the formulas below, the letter i represents a quantity associated with the general ith ellipse, $i = 1, 2, 3, \ldots, N$, where N is the number of ellipses involved. Each of these small ellipses is described in terms of its σ_{xi}, σ_{yi} and the angle θ_i between the x-axis and the arbitrarily selected reference axes (the w- and z-axes). The

TABLE 7.4 Contribution from Each Error Source

Source	u	Probability (%)
Weapon dispersion	3/40	37.89
Gun location	2/3	71.76
Observer location	3/4	48.19
Target location	1/3	42.26

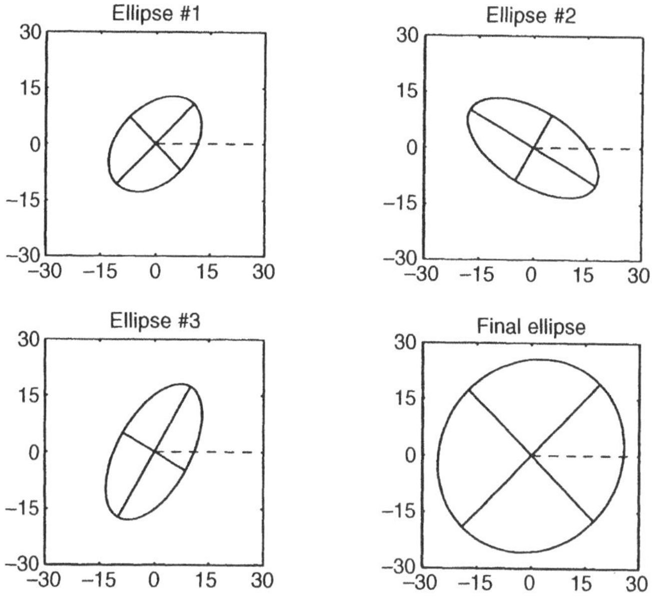

Figure 7.14 Input Error-Ellipses and Final Ellipse.

numerical values of the sigmas and orientation of the three input ellipses in Figure 7.14 are tabulated in Table 7.5.

TABLE 7.5 Parameters of Input Error-Ellipses

i	σ_{xi}(m)	σ_{yi}(m)	θ_i(°)
1	15	10	45°
2	10	20	60°
3	10	20	150°

To obtain the elements of the final ellipse, variances for each ellipse along the w- and z-axes are calculated. Since the w-, z-axes are not aligned with those of the individual ellipses, in general, an additional cross-product coefficient is required. These three elements (σ_{wi}^2, σ_{zi}^2, ρ_i for each ellipse) are then combined to obtain the corresponding elements in the final ellipse (σ_f^2, σ_f^2, ρ_f). The subscript f is used to denote parameters associated with the final ellipse. The terms σ_{wf}, σ_{zf}, ρ_f may then be converted to the σ_{xf}, σ_{yf} of the final ellipse along its major and minor axes.

First let us define the auxiliary functions to be used in the formulas. These are:

$$A_i = \frac{\cos^2\theta_i}{\sigma_{xi}^2} + \frac{\sin^2\theta_i}{\sigma_{yi}^2} \tag{7.8}$$

$$B_i = \frac{\cos^2\theta_i}{\sigma_{yi}^2} + \frac{\sin^2\theta_i}{\sigma_{xi}^2} \tag{7.9}$$

$$C_i = \cos\theta_i \sin\theta_i \left(\frac{1}{\sigma_{yi}^2} - \frac{1}{\sigma_{xi}^2}\right) \tag{7.10}$$

$$\rho_i = \frac{C_i}{\sqrt{A_i B_i}} \tag{7.11}$$

$$\sigma_{wi}^2 = \frac{1}{(1-\rho_i^2)A_i} \tag{7.12}$$

$$\sigma_{zi}^2 = \frac{1}{(1-\rho_i^2)B_i} \tag{7.13}$$

The last three equations give the cross-product coefficient and variances for each ellipse in the reference w- and z-axes. These are then combined according to the next three equations to obtain the corresponding elements for the final ellipse.

$$\sigma_{wf}^2 = \sum_{i=1}^{N} \sigma_{wi}^2 \tag{7.14}$$

$$\sigma_{zf}^2 = \sum_{i=1}^{N} \sigma_{zi}^2 \tag{7.15}$$

$$\rho_f = \frac{1}{\sigma_{wf}\,\sigma_{zf}} \sum_{i=1}^{N} (\rho_i\,\sigma_{wi}\,\sigma_{zi}) \tag{7.16}$$

We now have all the parameters needed for the final ellipse in terms of the w- and z-axes. To eliminate the cross-product coefficient ρ_f, and to obtain σ_{xf}, σ_{yf} along the major and minor axes of the final ellipse, we use the following formulas:

$$\sigma_{xf}^2 = \frac{1}{2}\left[\sigma_{wf}^2 + \sigma_{zf}^2 + \sqrt{(\sigma_{wf}^2 + \sigma_{zf}^2)^2 - 4\sigma_{wf}^2\,\sigma_{zf}^2(1-\rho_f^2)}\right] \tag{7.17}$$

$$\sigma_{yf}^2 = \frac{1}{2}\left[\sigma_{wf}^2 + \sigma_{zf}^2 - \sqrt{(\sigma_{wf}^2 + \sigma_{zf}^2)^2 - 4\sigma_{wf}^2\,\sigma_{zf}^2(1-\rho_f^2)}\right] \tag{7.18}$$

The orientation of the final ellipse relative to the w- and z-axes is given by the formula

$$\tan(2\theta_f) = \frac{2\rho_f\,\sigma_{wf}\,\sigma_{zf}}{\sigma_{wf}^2 - \sigma_{zf}^2} \tag{7.19}$$

The tedious computations associated with these equations are included in the M-file **novall.m** for MATLAB users. An input ASCII data file **nsd1.dat** containing the elements for the error-ellipses involved is needed for this program. For the example in Figure 7.14, the content of nsd1.dat consists of

```
15 10  45

10 20  60

10 20 150
```

The CEP (radius of $p = 50\%$ circle) corresponding to the final ellipse in this example is found, via the MATLAB command $R = noval1(0.5)$, to be 30.26 m. The program computes many other intermediate parameters in the process; these are

$$\sigma_{wf}^2 = 662.5000, \quad \sigma_{zf}^2 = 662.5000, \quad \rho_f = 0.0943,$$

$$\sigma_{xf} = 26.9258, \quad \sigma_{yf} = 24.4949, \quad \theta_f = 45°, \quad u = 0.9097$$

7.4 SUMMARY OF M-FILES

The M-files used or generated in this chapter are summarized in Table 7.6. Recall that the radius R and the normalized radius r are related by

TABLE 7.6 M-files Used in Chapter 7

File Name	MATLAB Command
bf1.m	$y = bf1(\lambda)$
bf2.m	$p = bf2(r, u)$
bf3.m	$y = bf3(r)$
bf4.m	$r = bf4(p, u)$
cf2.m	$p = cf2(r)$
cf3.m	$r = cf3(p)$
ang2p.m	$p = ang2p(\alpha, p_0)$
fsca.m	$fsca(\sigma_1, \sigma_2, \alpha)$
sem1.m	$p = sem1(R, \sigma_1, \sigma_2, \alpha)$
edcaud.m	$p = edcaud(R, \sigma_1, \sigma_2)$
sem2.m	$p = sem2(R, \sigma_1, \sigma_2, \alpha)$
noval.m	$p = noval(R)$
noval1.m	$R = noval1(p)$

$$r = R/\max(\sigma_x, \ \sigma_y)$$

The M-file **bf1.m** is used to implement the integrand of the integral defined in Equation (4.19) and repeated here

$$p = \frac{1}{u}\int_0^{R/\sigma_x} \lambda \exp\left[\frac{-\lambda^2(1+u^2)}{4u^2}\right] I_o\left[\frac{\lambda^2(1-u^2)}{4u^2}\right] d\lambda$$

whereas **bf2.m** evaluates the integral to determine the probability p. Both **sem2.m** and **noval.m** call bf2.m. The file **noval1.m** calls **bf4.m** which, in turn, calls **bf3.m**; **sem1.m** calls **edcaud.m**.

8

Risk Analysis

When a customer (buyer) purchases a high-technology system, it is necessary to know whether the system is as accurate as it is specified. A clearly defined set of test procedures, including pass/fail criterion, is required to determine if the system is acceptable to the customer. Because of the cost and time associated with each test, only a finite number of tests can be performed. The test results lead to estimates of the true characteristics of the system. Hence, a risk (based on using test statistics in decision making) is involved on the buyer's side in purchasing a bad system (true quality below specification), even though it passes the test. The seller also has his share of risk. That is, a good system (true quality equal to or better than the specification) can fail the test. The test pass/fail criterion therefore plays an important role in the risks of both buyer and seller. As more tests are performed, the test estimates become closer to the true system quality. In addition, given the specification, a conservative pass/fail criterion results in high buyer's risk, while a tight pass/fail criterion results in high seller's risk.

It is beneficial to both sides to understand how the pass/fail criterion is set [**16**]. Thus, the risks associated with the buyer as well as the seller can be made both reasonable and acceptable.

This chapter outlines the procedures for determining the pass/fail criterion and buyer's risk for a given seller's risk utilizing an associated specification supplied by the buyer.

We will assume that the system under consideration is an Inertial Navigation Unit (INU) and that the particular parameter of concern is the CEP of the INU. The true INU CEP can be denoted as $r_{0.5}$ and the computed/estimated CEP, from an ensemble of N tests, as $\hat{r}_{0.5}$.

We formulate the problem in the fashion of hypothesis testing, with the hypotheses stated in terms of the true population CEP, $r_{0.5}$. The **null hypothesis** is

$$H_0: r_{0.5} = C_0$$

that is, the true CEP of the INU meets the specification, C_0, and the **alternate hypothesis** is

$$H_1: r_{0.5} = C_1 > C_0$$

that is, the true CEP of the INU exceeds the specification, C_0. Note that the symbol C_1 is used for $r_{0.5}$, when $r_{0.5} > C_0$.

Given a specification C_0, the **seller's risk** α is the conditional probability of rejecting (based on test result) a good INU (true CEP $r_{0.5} = C_0$). Assume that an ensemble of N tests produces a computed CEP exceeding the pass/fail criterion ($\hat{r}_{0.5} \geq L$). That is,

$$\alpha = P\{\hat{r}_{0.5} \geq L | r_{0.5} = C_0\}$$

The **buyer's risk** β is the conditional probability of purchasing (based on test result) a bad INU (true CEP $r_{0.5} = C_1 > C_0$), if an ensemble of N tests shows that the computed CEP is below the set pass/fail criterion ($\hat{r}_{0.5} < L$). Thus,

$$\beta = P\{\hat{r}_{0.5} < L | r_{0.5} = C_1\}$$

The basic assumptions made for risk analysis are as follows:

1. Random variables (position errors in the east and north directions) are assumed to be independent, zero mean, and with the same normal distribution; that is, $\mathbf{x} \sim N(0, \sigma^2)$ and $\mathbf{y} \sim N(0, \sigma^2)$.
2. Population standard deviation σ is to be estimated with $2N$ independent samples from N tests, and the true INU CEP ($r_{0.5}$) is related to σ by $r_{0.5} = 1.1774 \cdot \sigma$ (see Chapter 4).
3. Position error rates (CEP rates) are treated as position errors (CEPs) that occur at one hour. Hence, the analysis is based on position error, while the results are interpreted as position error per hour, or error rate.
4. Estimated CEP is computed from

$$\hat{r}_{0.5} = 1.1774 \cdot S_{2N}, \quad \text{with } S_{2N}^2 = \frac{1}{2N} \sum_{i=1}^{N} (x_i^2 + y_i^2)$$

Section 8.1 of this chapter further defines the notation. The following three questions are answered by the analysis in Sections 8.2, 8.3, and 8.4, respectively.

- Given the specification C_0, the pass/fail criterion, and the number of tests N, what is the seller's risk, α?
- Given the seller's risk α and specification C_0, what is the pass/fail criterion L, or, in terms of C_0, what is $K_\alpha (= L/C_0)$?
- Given K_α and C_1 (or $\lambda = C_1/C_0$), what is the buyer's risk β?

8.1 DEFINITION OF NOTATION

The notation used specifically in this chapter is defined as follows.

C_0 = Customer Specified Acceptance Requirement

C_1 = a symbol representing the true INU CEP when it exceeds C_0

$\lambda = C_1/C_0 > 1$

L = pass/fail criterion = $K_\alpha C_0$

K_α = normalized pass/fail criterion = L/C_0

x = INU east position error

y = INU north position error

σ = true INU standard deviation of x or y position error

N = number of tests

S_{2N} = sample standard deviation from N tests

$r_{0.5}$ = true INU Circular Error Probable

$\hat{r}_{0.5}$ = estimated Circular Error Probable from an ensemble of N tests

χ_m^2 = chi-square distribution with m degrees of freedom

$\chi_{m,a}^2$ = the point to the right of which a random variable with χ_m^2 distribution has an area a under the PDF

$\chi_{m,1-b}^2$ = the point to the left of which a random variable with χ_m^2 distribution has an area b under the PDF

P = probability of the occurrence of some event

$P\{A|B\}$ = P\{event A under the condition that event B has occurred\}

α = seller's risk = $P\{\hat{r}_{0.5} \geq L | r_{0.5} = C_0\}$

β = buyer's risk = $P\{\hat{r}_{0.5} < L | r_{0.5} = C_1\}$

Notice that we have expressed L and C_1 in terms of the specification C_0 as $L = K_\alpha C_0$ and $C_1 = \lambda C_0$, respectively.

The CEP concept is introduced in Chapter 4, and the basic properties of the chi-square PDF are described in Chapter 6 as well as in Appendix A.

The geometric meaning of $\chi_{m,a}^2$ and $\chi_{m,1-b}^2$ is shown in Figure 8.1. The area to the right of the point $\chi_{m,a}^2$, and under the χ_m^2 curve is a. Similarly, the area to the left of the point $\chi_{m,1-b}^2$, and under the χ_m^2 curve is b.

8.2 SELLER'S RISK

According to Theorem 2.2.1 in Chapter 2, we know that the random variable $2NS_{2N}^2/\sigma^2$ has a χ^2 distribution with $(2N - 1)$ degrees of freedom; the new random

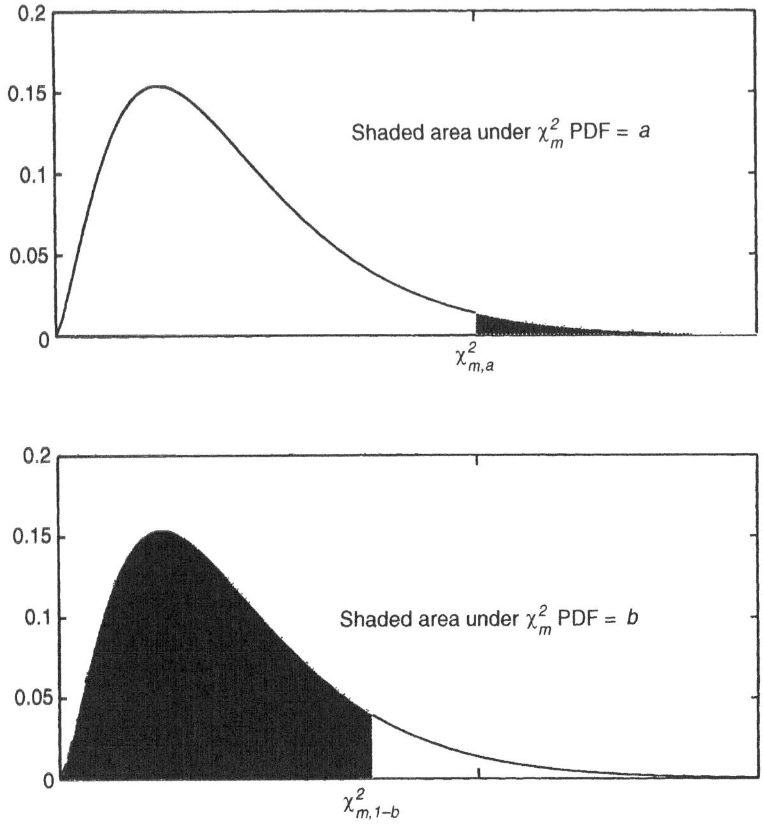

Figure 8.1 Geometric Meaning of $\chi^2_{m,a}$ and $\chi^2_{m,1-b}$.

variable $2N(1.1774S_{2N})^2/(1.1774\sigma)^2$ (obtained by multiplying 1.1774^2 to both the numerator and the denominator of the old random variable $2NS^2_{2N}/\sigma^2$)

$$2N(1.1774S_{2N})^2/(1.1774\sigma)^2 = 2N(\hat{r}_{0.5})^2/r^2_{0.5}$$

also possesses the same χ^2 distribution with $(2N - 1)$ degrees of freedom.

The relationship between the pass/fail criterion L (or its normalized version K_α) and the seller's risk α can be derived from

$$\begin{aligned}\alpha &= P\{\hat{r}_{0.5} \geq L | r_{0.5} = C_0\} \\ &= P\{2N(\hat{r}_{0.5}/r_{0.5})^2 \geq 2N(L/r_{0.5})^2 | r_{0.5} = C_0\} \\ &= P\{\chi^2_{2N-1} \geq 2N(L/C_0)^2\}\end{aligned} \tag{8.1}$$

Or using the convention defined in Figure 8.1, we have

$$\alpha = P\{\chi^2_{2N-1} \geq \chi^2_{2N-1,\alpha}\} \tag{8.2}$$

with $\chi^2_{2N-1,\alpha} = 2N(L/C_0)^2$.

Thus, determination of the seller's risk can be classified as a direct problem:

Given C_0, L, N, *determine the probability* α *from Equation (8.2).*

The solution process for the direct problem is implemented in **cln2a.m**.

8.3 THE PASS/FAIL CRITERION

Finding the pass/fail criterion L, when we know the seller's risk α, the specification C_0, and the number of tests N, corresponds to the inverse problem:

Find $\chi^2_{2N-1,\alpha}$, *given the seller's risk* α *and the number of tests* N.

The pass/fail criterion can then be obtained from

$$L = \sqrt{\frac{\chi^2_{2N-1,\alpha}}{2N}} C_0 \tag{8.3}$$

or the normalized pass/fail criterion from

$$K_\alpha = L/C_0 = \sqrt{\frac{\chi^2_{2N-1,\alpha}}{2N}} \tag{8.4}$$

Notice that K_α is a function of α and N. The M-file **an2ka.m** computes K_α via the MATLAB command $K_\alpha = an2ka(\alpha, N)$.

Since $\chi^2_{2N-1,\alpha} = 2N(L/C_0)^2$, we see from Equation (8.2) that the seller's risk α increases when C_0 decreases. Also, α decreases when N or L increases. Figure 8.2 shows the dependency of K_α with N and the parameter α (seller's risk).

Two facts are obvious from Figure 8.2:

1. For fixed N, K_α increases as α decreases.
2. For fixed α, K_α approaches 1 as N increases toward ∞.

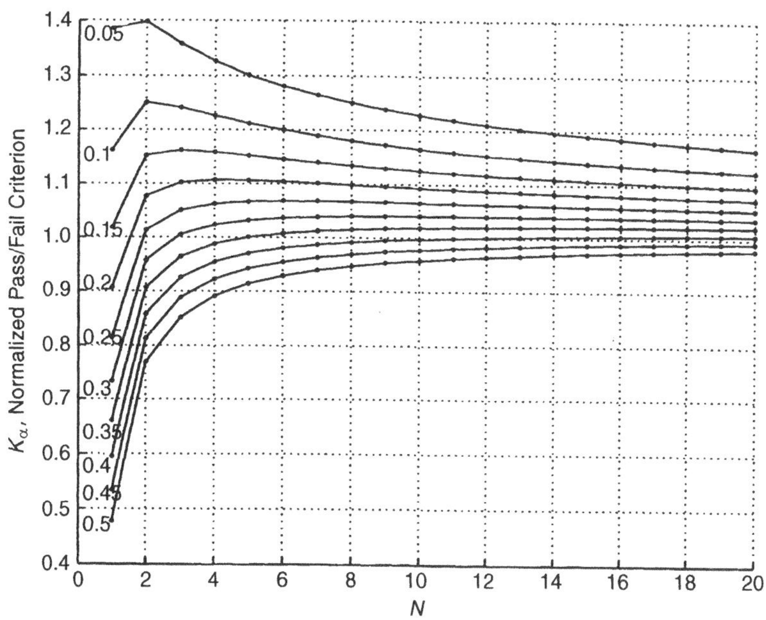

Figure 8.2 Normalized Pass/Fail Criterion K_α as a Function of the Number of Tests N, with Seller's Risk α as Parameter.

8.4 BUYER'S RISK

Once the pass/fail criterion L is set, the corresponding buyer's risk β can be derived from

$$\begin{aligned}\beta &= P\{\hat{r}_{0.5} < L | r_{0.5} = C_1\} \\ &= P\{2N(\hat{r}_{0.5}/r_{0.5})^2 < 2N(L/r_{0.5})^2 | r_{0.5} = C_1\} \\ &= P\{\chi^2_{2N-1} < 2N(L/C_1)^2\} \\ &= P\{\chi^2_{2N-1} < 2N[L/(\lambda C_0)]^2\} \\ &= P\{\chi^2_{2N-1} < 2N(K_\alpha/\lambda)^2\} \end{aligned} \tag{8.5}$$

Hence,

$$\beta = P\{\chi^2_{2N-1} < \chi^2_{2N-1,1-\beta}\} \tag{8.6}$$

with

$$\chi^2_{2N-1,1-\beta} = 2N\left(\frac{K_\alpha}{\lambda}\right)^2 \tag{8.7}$$

which depends on N, K_α, and λ.

Given α, N, we can determine K_α from Equation (8.4) of the previous section. Thus, when α, N, λ are given, we know N, K_α, λ, and thus $\chi^2_{2N-1,1-\beta}$. The M-file **anl2b.m** carries out all the necessary computations to obtain β via the MATLAB command $\beta = anl2b(\alpha, N, \lambda)$.

Substituting K_α of Equation (8.4) into Equation (8.7), we have

$$\chi^2_{2N-1,1-\beta} = \frac{\chi^2_{2N-1,\alpha}}{\lambda^2}$$

Therefore, λ, which equals the ratio C_1/C_0, also satisfies

$$\lambda = \sqrt{\frac{\chi^2_{2N-1,\alpha}}{\chi^2_{2N-1,1-\beta}}} \tag{8.8}$$

Figure 8.3 shows how β varies with K_α, for a fixed $N = 8$, using λ as parameter. Two facts are obvious:

1. For fixed N, regardless of λ, buyer's risk increases as K_α increases.
2. For fixed K_α, buyer's risk decreases as λ increases.

Figure 8.4 shows the relationship between the normalized pass/fail criterion K_α and N, the number of tests, using seller's risk as a parameter. From this figure, we see that

1. For fixed N, K_α decreases as α increases.
2. For fixed α, K_α decreases as N increases.

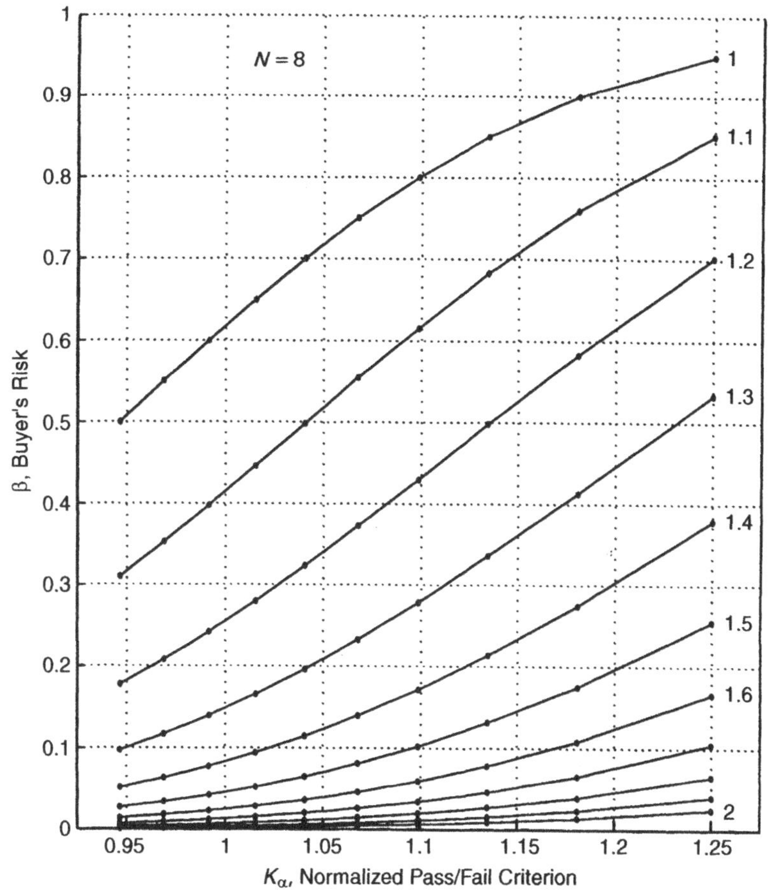

Figure 8.3 Buyer's Risk versus K_α, for $N = 8$, with λ as Parameter.

8.5 A PRACTICAL EXAMPLE

Suppose we are given

(1) $N = 8 =$ the number of tests
(2) $C_0 = 0.8 =$ the CEP specification
(3) $L = 1.0 =$ the CEP pass/fail criterion

The seller's risk

$$\alpha = P\{\hat{r}_{0.5} \geq L | r_{0.5} = C_0\}$$

and the buyer's risk

$$\beta = P\{\hat{r}_{0.5} < L | r_{0.5} = C_1 > C_0\}$$

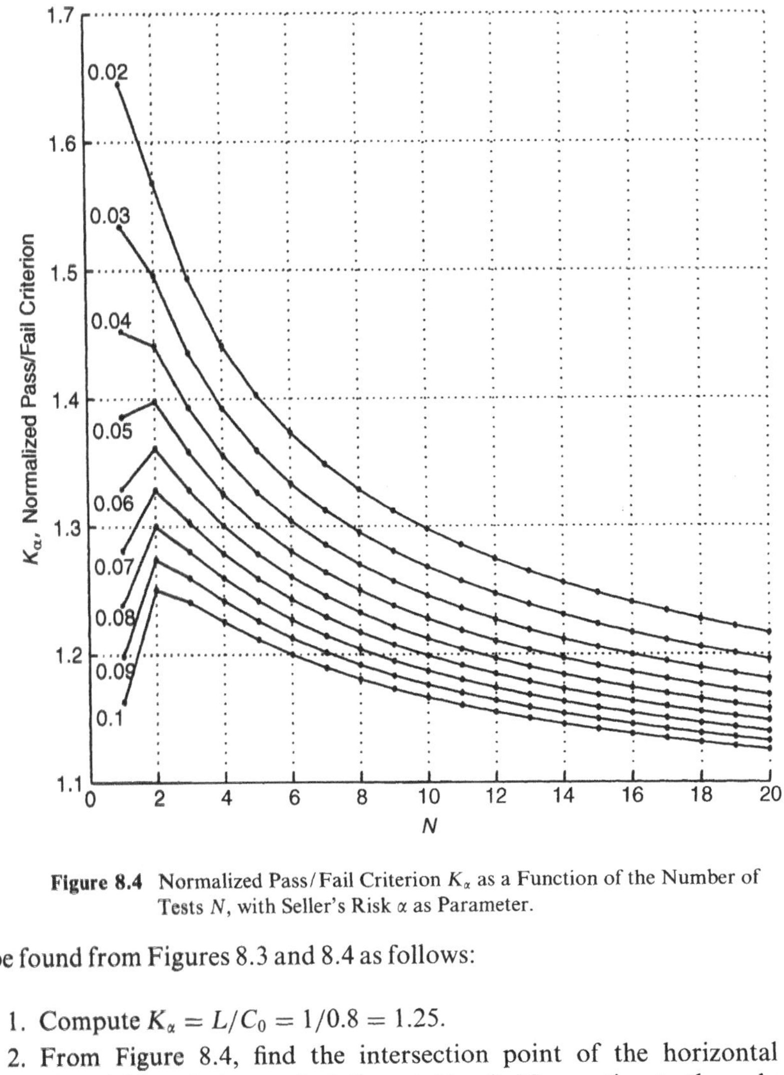

Figure 8.4 Normalized Pass/Fail Criterion K_α as a Function of the Number of Tests N, with Seller's Risk α as Parameter.

can be found from Figures 8.3 and 8.4 as follows:

1. Compute $K_\alpha = L/C_0 = 1/0.8 = 1.25$.
2. From Figure 8.4, find the intersection point of the horizontal line at $K_\alpha = 1.25$ and the vertical line at $N = 8$. Now estimate the value of the parameter α associated with the curve passing through that point. This produces the seller's risk at $\alpha \approx 5\%$.
3. Assume $C_1 = 1.12$ so that $\lambda = C_1/C_0 = 1.12/0.8 = 1.4$.
4. From Figure 8.3, find the intersection point of the curve associated with $\lambda = 1.4$ and the vertical line at $K_\alpha = 1.25$. The ordinate of this intersection point is the buyer's risk, $\beta \approx 38\%$. Note that β is dependent on the choice of λ and hence the value of C_1.

To avoid visual interpolation with figures, we can use the MATLAB M-file **cln2a.m** and **cln2b.m** to find the seller's risk and buyer's risk, α, β, respectively. Employing the values of C_0, L, and N shown above, we enter the command

```
alpha = cln2a(0.8, 1, 8)
```

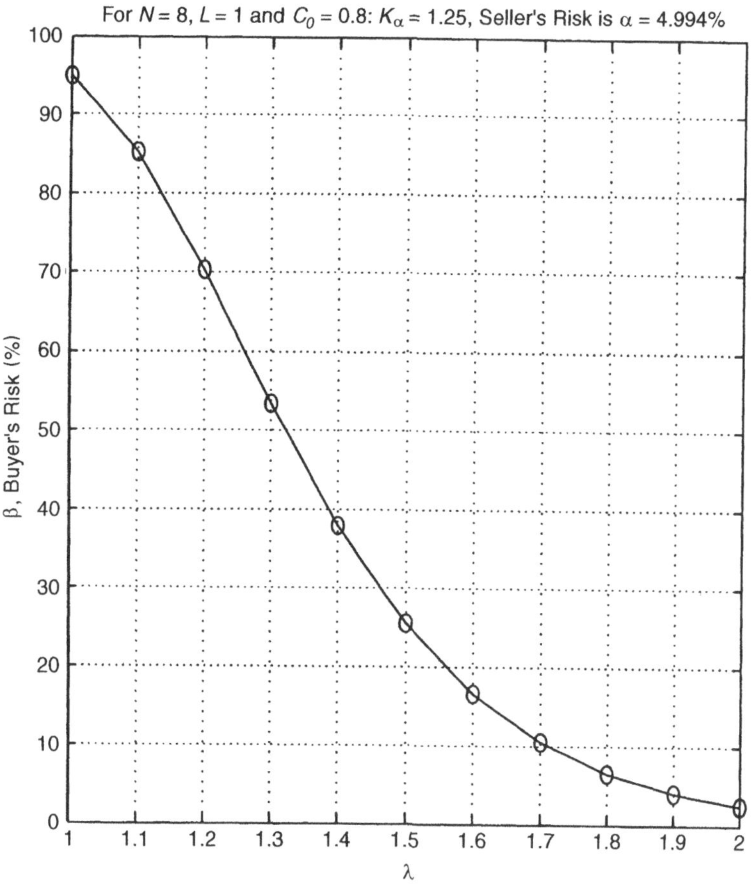

Figure 8.5 Buyer's Risk for $K_\alpha = 1.25$ and $N = 8$, with λ as Parameter.

and find $\alpha = 4.9943\%$. The general form of this command is $\alpha = cln2a(C_0, L, N)$.

With the particular value of $\lambda = 1.4$ (assume $C_1 = 1.12$), we can obtain the buyer's risk $\beta = 37.8794\%$ via the command

```
beta = cln2b(0.8, 1, 8, 1.4)
```

Thus, the values extracted from Figures 8.3 and 8.4 for β and α, respectively, are good approximations.

If we let λ vary from 1 to 2, we will obtain the buyer's risks as displayed in Figure 8.5.

8.6 GENERALIZATION

In the preceding sections we considered the seller's risk, assuming $r_{0.5} = C_0$. A reasonable question is, "What is the chance that the system passes the test but has a quality better than the specification?" That is, find the probability $P\{\hat{r}_{0.5} \geq L \mid r_{0.5} < C_0\}$.

Thus, the more general problem should be formulated to include inequalities for $r_{0.5}$ such as

$$\alpha = P\{\hat{r}_{0.5} \geq L | r_{0.5} \leq C_0\} \tag{8.9}$$

$$\beta = P\{\hat{r}_{0.5} < L | r_{0.5} > C_0\} \tag{8.10}$$

Let us first define $r_{0.5} = K_0 C_0$, with $K_0 \leq 1$ in the expression for α, and $r_{0.5} = K_1 C_0$, with $K_1 > 1$ in the expression for β. Also, if we define $\Lambda = K_1/K_0$, then we can obtain the expressions

$$K_\alpha = \sqrt{\frac{\chi^2_{2N-1,\alpha}}{2N}} K_0 \tag{8.11}$$

$$\chi^2_{2N-1,1-\beta} = 2N\left(\frac{K_\alpha}{K_1}\right)^2 \tag{8.12}$$

Since $K_0 \leq 1$, the normalized pass/fail criterion K_α will be scaled down when compared with Equation (8.4). By treating $K_1 C_0$ as C_1 in the preceding section, we have $K_1 = \lambda$. Therefore, $\chi^2_{2N-1,1-\beta}$ will be smaller on comparing Equations (8.7) with (8.12), resulting in a lower buyer's risk.

Note that with $\Lambda = K_1/K_0$, and the last two equations, Λ satisfies

$$\Lambda = \sqrt{\frac{\chi^2_{2N-1,\alpha}}{\chi^2_{2N-1,1-\beta}}} \tag{8.13}$$

just as with λ in Equation (8.8) of Section 8.3.

8.7 SUMMARY OF M-FILES

The M-files used or generated in this chapter are summarized in Table 8.1.

Notice, however, that the direct problems (given R, find $p = \int_0^R g(t)\,dt$, with $g(t)$ being the PDF of $\mathbf{t} \sim \chi^2_n$) and the inverse problems (given p, find R) involved in risk analysis are solved by an interpolation technique utilizing auxiliary files. The details are explained as follows.

TABLE 8.1 M-files Used in Chapter 8

File Name	MATLAB Command
cdfchi2.m	$p = cdfchi2(x, N), \quad N \leq 40$
ichi2.m	$x = ichi2(p, N), \quad N \leq 40$
an2ka.m	$K_\alpha = an2ka(\alpha, N)$
anl2b.m	$\beta = anl2b(\alpha, N, \lambda)$
cln2a.m	$\alpha = cln2a(C_0, L, N)$
cln2b.m	$\beta = cln2b(C_0, L, N, \lambda)$

The author has created 40 files named p2x1.mat, p2x2.mat, ..., p2x40.mat, corresponding to the positive integer $n = 1, 2, \ldots, 40$. Each of these files contains two vectors $\vec{R} = [R_1, R_2, \ldots, R_{k(n)}]'$ and $\vec{p} = [p_1, p_2, \ldots, p_{k(n)}]'$ such that

$$p_j = \int_0^{R_j} g(t)\,dt, \quad \text{for } j = 1, 2, \ldots, k(n)$$

with $R_1 = 0$, $p_1 = 0$. The components in $\vec{R}$ are equally spaced at an interval of 0.1. Thus, the largest component $R_{k(n)}$ is determined by the degree of freedom n and the condition

$$\int_0^{R_k(n)} g(t)\,dt \cong 0.9999$$

For direct problems where R is given, we use the MATLAB built-in interpolation command (**interp1.m**), and find p from $p = interp1(\vec{R}, \vec{p}, R)$. For inverse problems where p is given, we find R from $R = interp1(\vec{p}, \vec{R}, p)$.

The two M-files **cdfchi2.m** and **ichi2.m** perform the interpolation operation using one of the 40 files p2x1.mat, p2x2.mat, p2x3.mat,..., p2x40.mat.

A

Probability Density Functions

A.1 RELATIONSHIP BETWEEN PDF AND CDF

The probability density function $g(t)$ of a random variable $\mathbf{t}$ is related to the cumulative distribution function $G(t)$ via

$$G(t) = \int_{-\infty}^{t} g(u)\,du \tag{A.1}$$

or

$$g(t) = dG(t)/dt \tag{A.2}$$

Since $g(t) \geq 0$ for all t, $G(t)$ is a monotonically nondecreasing function. See **[15]** for a detailed description.

A.2 PDFS USED IN THIS BOOK

A.2.1 Normal Distribution

A normally distributed random variable $\mathbf{t}$ with mean μ and standard deviation σ has the PDF

$$g(t) = \frac{1}{\sqrt{2\pi}\sigma} \exp[-(t-\mu)^2/(2\sigma)] \triangleq N(\mu, \sigma^2), \quad \text{for } t \in (-\infty, \infty)$$

and is abbreviated as $\mathbf{t} \sim N(\mu, \sigma^2)$. Its kth central moment is

$$E[(\mathbf{t}-\mu)^k] = \begin{cases} 1 \cdot 3 \cdot 5 \cdots (k-1)\sigma^k, & \text{for even } k \\ 0, & \text{for odd } k \end{cases}$$

When $\mu = 0$ and $\sigma = 1$, we have the standard normal distribution. Its PDF becomes

$$g(t) = \frac{1}{\sqrt{2\pi}} \exp(-t^2/2) \triangleq N(0, 1), \quad \text{for } t \in (-\infty, \infty)$$

and is abbreviated as $\mathbf{t} \sim N(0, 1)$. The kth absolute moment of the standard normal distribution is

$$E[|\mathbf{t}|^k] = 2^{(k/2)}\left(\frac{k-1}{2}\right)!/\sqrt{\pi}$$

A.2.2 Chi-Square Distribution

A chi-square distributed random variable $\mathbf{t}$ with n degrees of freedom has the PDF

$$g_n(t) = \frac{1}{\Gamma(n/2)2^{n/2}} t^{(n/2)-1} \exp(-t/2) \triangleq \chi_n^2, \quad \text{for } t \in [0, \infty)$$

and is abbreviated as $\mathbf{t} \sim \chi_n^2$. The symbol $\Gamma(x)$ represents the gamma function evaluated at x. The mean and standard deviation of the random variable $\mathbf{t}$ are n and $2n$, respectively.

Let $G_n(x) = \int_0^x g_n(t)\,dt$ be the CDF. We can apply the method of integration by parts to obtain the recurrence formula

$$G_n(x) = \frac{x^{n/2} \exp(-x/2)}{\Gamma(n/2+1)2^{n/2}} + G_{n+2}(x) \tag{A.3}$$

This formula is useful in circumventing the singularity of $g_n(t)$ at $t = 0$, for $n < 2$, when evaluating the CDF $G_n(x)$ numerically.

A.2.3 Student-*t* Distribution

A student-t distributed random variable $\mathbf{t}$ with n degrees of freedom is given as

$$g(t) = \frac{\Gamma[(n+1)/2]}{\Gamma[n/2]\sqrt{n\pi}} \frac{1}{[1+(t^2/n)^{(n+1)/2}]}, \quad \text{for } t \in (-\infty, \infty)$$

where the symbol $\Gamma(x)$ denotes the gamma function evaluated at x. It has a mean of zero and a standard deviation of $\sqrt{n/(n-2)}$, for $n > 2$.

A.2.4 Rayleigh Distribution

A random variable $\mathbf{t}$ with Rayleigh distribution has the PDF

$$g(t) = \frac{t}{\sigma^2} \exp\left[\frac{-t^2}{2\sigma^2}\right], \quad \text{for } t \in [0, \infty)$$

with a mean of $\sqrt{(\pi/2)}\sigma$ and a standard deviation of $\sqrt{(2-\pi/2)}\sigma$.

A.2.5 Maxwell Distribution

A random variable **t** with Maxwell distribution has the PDF

$$g(t) = \sqrt{\frac{2}{\pi}} \frac{t^2}{\sigma^3} \exp\left[\frac{-t^2}{2\sigma^2}\right], \quad \text{for } t \in [0, \infty)$$

with a mean of $2\sqrt{(2/\pi)}\,\sigma$ and a standard deviation of $\sqrt{(3 - 8/\pi)}\,\sigma$.

A.2.6 Cauchy Distribution

A random variable **t** with Cauchy distribution has the PDF

$$g(t) = \frac{a/\pi}{a^2 + t^2}, \quad \text{for all real } t$$

Neither the mean nor the standard deviation of this distribution is defined.

A.2.7 Uniform Distribution

A random variable **t** with uniform distribution has the PDF

$$g(t) = \frac{1}{b - a}, \quad \text{for } t \in [a, b]$$

with a mean of $(a + b)/2$ and a standard deviation of $(b - a)/\sqrt{12}$.

A.3 CENTRAL LIMIT THEOREM

If a random variable **z** is expressed as the sum of n independent random variables (satisfying certain conditions that hold in most applications), then the PDF of **z** approaches the standard normal distribution as n becomes sufficiently large. This is known as the central limit theorem.

Theorem A.3.1 Let $\mathbf{x}_1, \mathbf{x}_2, \ldots, \mathbf{x}_n$ be a set of independent random variables with $E[\mathbf{x}_i] = \mu_i$ and $var[\mathbf{x}_i] = E[(\mathbf{x}_i - \mu_i)^2] = \sigma_i^2$, for $i = 1, 2, \ldots, n$. Let $\mathbf{x} = \mathbf{x}_1 + \mathbf{x}_2 + \cdots + \mathbf{x}_n$. Then the random variable

$$\mathbf{z}_n = \frac{\mathbf{x} - \sum_{i=1}^{n} \mu_i}{\sqrt{\sum_{i=1}^{n} \sigma_i^2}}$$

has the standard normal or $N(0, 1)$ distribution, as $n \to \infty$.

A.4 GENERATION OF STANDARD NORMAL RANDOM VARIABLE

As an application of the central limit theorem, let us take 12 independent random variables, each with the same uniform distribution from [0, 1].

Since $a = 0$ and $b = 1$, each of the 12 random variables has a mean of $E[\mathbf{x}_i] = (a + b)/2 = 1/2$, and a variance of $var[\mathbf{x}_i] = E[(\mathbf{x}_i - \mu_i)^2] = \sigma_i^2 = (b - a)/\sqrt{12} = 1/\sqrt{12}$, for $i = 1, 2, \ldots, 12$.

Thus, with $\mathbf{x} = \mathbf{x}_1 + \mathbf{x}_2 + \cdots + \mathbf{x}_{12}$, the random variable

$$\mathbf{z} = \frac{\mathbf{x} - \sum_{i=1}^{12} \mu_i}{\sqrt{\sum_{i=1}^{12} \sigma_i^2}} = \mathbf{x} - 6$$

has a PDF approximating a standard normal PDF.

In other words, if we take the sum of 12 uniformly distributed random numbers from the [0, 1] interval and then subtract 6 from this sum, the resulting random number will have a PDF very close to the $N(0, 1)$ PDF.

B

Method of Confidence Intervals

Using the definition given in Papoulis **[15]**, we can say that the interval (θ_1, θ_2) is a γ-confidence interval of a population parameter θ if

$$\text{Probability}\{\theta_1 < \theta < \theta_2\} = \gamma$$

where γ is a specified constant. It is called the **confidence coefficient**. The difference $\alpha = 1 - \gamma$ is the **confidence level** of the estimate. The statistics θ_1 and θ_2 are called the **confidence limits**.

B.1 CONFIDENCE INTERVAL AND CONFIDENCE LIMITS

Suppose we take a sample of N measurements of a normally distributed random variable $\mathbf{x}$ that has a variance known to be equal to 16. Suppose further that the mean computed from this sample is 30. Hence, $\sigma^2 = 16$ and $m = 30$. The problem is to estimate the population mean μ using an interval of values for x. It is a well-known fact that the random variable $\mathbf{m}$ has a normal PDF with $\mu_m = \mu$ and $\sigma_m = \sigma/\sqrt{N}$; that is, $\mathbf{m} \sim N(\mu, \sigma^2/N)$. As an example, if $N = 100$, then $\sigma_m = 4/\sqrt{100} = 0.4$. Thus, we can write

$$\text{Probability}\{-2\sigma_m < \mathbf{m} - \mu < 2\sigma_m\} = 0.9544 \tag{B.1}$$

This equation is equivalent to

$$\text{Probability}\{\mu - 2\sigma_m < \mathbf{m} < \mu + 2\sigma_m\} = 0.9544 \tag{B.2}$$

which means that, for a large number of experiments of this type, 95.44% of the trials will generate a value of $\mathbf{m}$ that falls within the interval $(\mu - 0.8, \mu + 0.8)$. In addition, after some algebraic manipulation,

$$\text{Probability}\{\mathbf{m} - 2\sigma_m < \mu < \mathbf{m} + 2\sigma_m\} = 0.9544 \tag{B.3}$$

which means that, for a large number of experiments of this type, 95.44% of the trials will yield an interval ($\mathbf{m} - 2\sigma_m$, $\mathbf{m} + 2\sigma_m$) that includes the unknown population mean μ.

In common practice, the experiment of taking 100 samples is made only once; these samples are used to compute the sample mean m. Hence, there is only one value for this random variable. The interval $(m - 2\sigma_m, m + 2\sigma_m) = (30 - 0.8, 30 + 0.8) = (29.2, 30.8)$ is called a 95.44% confidence interval for μ. The end points of a confidence interval associated with a parameter are called the confidence limits for the parameter.

B.2 DETERMINATION OF γ-CONFIDENCE INTERVAL

In general, these are the steps for finding the confidence interval:

- Find a random variable (call it $\mathbf{z}$) that involves the desired parameter θ but whose distribution does not depend on any known parameters. In the example above, $\theta = \mu$, $\mathbf{z} = \mathbf{m} - \mu$.
- Choose two numbers z_1 and z_2 such that Probability$\{z_1 < \mathbf{z} < z_2\} = \gamma$, where γ is the desired confidence coefficient, such as 95.44%.
- Transform the probability statement to Probability$\{\theta_1 < \theta < \theta_2\} = \gamma$.
- (θ_1, θ_2) is the p-confidence interval for the parameter θ.

Usually, a confidence level α is given instead of specifying the confidence coefficient γ. The relationship $\gamma = 1 - \alpha$ should make it easy to change from one condition to the other.

C

Function of N Random Variables

In many applications where several random variables are combined to form a new random variable, it is important to know the mean and variance of this new random variable. We summarize the major results for linear functions in Sections C.1 and C.2 of this appendix. The concepts of independence and correlation between two random variables are defined and illustrated in Section C.3. In Section C.4, we show that the PDF of the sum of two jointly normal random variables is also normally distributed and that the PDF of the quotient of two jointly normal random variables is Cauchy distributed.

C.1 LINEAR COMBINATION OF N INDEPENDENT RANDOM VARIABLES

Let $\{\mathbf{x}_1, \mathbf{x}_2, \mathbf{x}_3, \ldots, \mathbf{x}_N\}$ be a collection of N independent random variables. If $\mathbf{z}$ is a linear function of these variables,

$$\mathbf{z} = a_1\mathbf{x}_1 + a_2\mathbf{x}_2 + \cdots + a_N\mathbf{x}_N \tag{C.1}$$

then the mean and variance of the newly created random variable $\mathbf{z}$ are

$$\mu_z = E[\mathbf{z}] = a_1\mu_{x_1} + a_2\mu_{x_2} + \cdots + a_N\mu_{x_N} \tag{C.2}$$

and

$$\sigma_z^2 = E[(\mathbf{z} - m_z)^2] = a_1^2\sigma_{x_1}^2 + a_2^2\sigma_{x_2}^2 + \cdots + a_N^2\sigma_{x_N}^2 \tag{C.3}$$

respectively.

C.1.1 Sum of *N* Random Variables

When all coefficients $a_i = 1,\ i = 1, 2, \ldots, N$, $\mathbf{z}$ is simply the sum of all the N random variables $\mathbf{x}_i,\ i = 1, 2, \ldots, N$. In this case,

$$\mu_z = \sum_{i=1}^{N} \mu_{x_i} \quad \text{and} \quad \sigma_z^2 = \sum_{i=1}^{N} \sigma_{x_i}^2$$

Suppose further that $\mu_{x_i} = \mu_x,\ \sigma_{x_i}^2 = \sigma_x^2,\ i = 1, 2, \ldots, N$, then the mean and variance of $\mathbf{z}$ are reduced to

$$\mu_z = N\mu_x \quad \text{and} \quad \sigma_z^2 = N\sigma_x^2$$

C.1.2 Average of *N* Random Variables

When all coefficients $a_i = 1/N,\ i = 1, 2, \ldots, N$, then $\mathbf{z}$ represents the average of these N random variables $\mathbf{x}_i,\ i = 1, 2, \ldots, N$. In this case,

$$\mu_z = (1/N) \sum_{i=1}^{N} \mu_{x_i} \quad \text{and} \quad \sigma_z^2 = (1/N^2) \sum_{i=1}^{N} \sigma_{x_i}^2$$

Suppose further that $\mu_{x_i} = \mu_x,\ \sigma_{x_i}^2 = \sigma_x^2$, for $i = 1, 2, \ldots, N$, then the mean and variance of $\mathbf{z}$ are simplified to

$$\mu_z = \mu_x \quad \text{and} \quad \sigma_z^2 = (1/N)\sigma_x^2$$

C.1.3 Difference of Two Random Variables

When $N = 2$ and $a_1 = 1,\ a_2 = -1$, $\mathbf{w}$ is the difference of random variables $\mathbf{x}_1$ and $\mathbf{x}_2$. In this case,

$$\mu_x = \mu_{x_1} - \mu_{x_2} \quad \text{and} \quad \sigma_x^2 = \sigma_{x_1}^2 + \sigma_{x_2}^2$$

Suppose further that $\mu_{x_i} = \mu_x,\ \sigma_{x_i}^2 = \sigma_x^2$, for $i = 1, 2$, then the mean and variance of $\mathbf{z}$ become

$$\mu_x = 0 \quad \text{and} \quad \sigma_x^2 = 2\sigma_x^2$$

Even though the mean values are different for the sum $(\mathbf{x}_1 + \mathbf{x}_2)$ and the difference $(\mathbf{x}_1 - \mathbf{x}_2)$, their variances are the same, provided $\mathbf{x}_1$ and $\mathbf{x}_2$ are independent random variables.

C.2 PRODUCT OF TWO RANDOM VARIABLES

Let $\mathbf{x}_1, \mathbf{x}_2$ be two independent random variables. Define the new random variable $\mathbf{z} = \mathbf{x}_1\mathbf{x}_2$. The mean of this product random variable $\mathbf{z}$ is

$$\mu_z = E[\mathbf{z}] = E[\mathbf{x}_1]E[\mathbf{x}_2] = \mu_1\mu_2 \tag{C.4}$$

and the variance of **z** is

$$\begin{aligned}\sigma_z^2 &= E[(\mathbf{z} - \mu_z)^2] = E[(\mathbf{x}_1\mathbf{x}_2 - \mu_1\mu_2)^2] \\ &= E[\mathbf{x}_1^2\mathbf{x}_2^2 - 2\mu_1\mu_2\mathbf{x}_1\mathbf{x}_2 + \mu_1^2\mu_2^2] \\ &= E[\mathbf{x}_1^2\mathbf{x}_2^2] - \mu_1^2\mu_2^2 \\ &= \mu_2^2\sigma_{x_1}^2 + \mu_1^2\sigma_{x_2}^2 \end{aligned} \tag{C.5}$$

C.3 SUM AND DIFFERENCE OF TWO CORRELATED RANDOM VARIABLES

The previous sections deal with random variables that are pairwise independent. Note that independence implies lack of correlation but that the converse is, in general, not true, as described in the next section.

If the random variables $\mathbf{x}_1$, $\mathbf{x}_2$ are correlated, then their sum $\mathbf{z} = \mathbf{x}_1 + \mathbf{x}_2$ has a mean that is exactly the same as in the independent case,

$$\mu_z = \mu_{x_1} + \mu_{x_2}$$

but the variance of **z** takes the form

$$\sigma_z^2 = \sigma_{x_1}^2 + \sigma_{x_2}^2 + 2\sigma_{x_1x_2} = \sigma_{x_1}^2 + \sigma_{x_2}^2 + 2\rho_{x_1x_2}\sigma_{x_1}\sigma_{x_2} \tag{C.6}$$

where $\sigma_{x_1x_2}$ is the **covariance of** $\mathbf{x}_1$ and $\mathbf{x}_2$, σ_{x_i} is the variance of $\mathbf{x}_i$, and $\rho_{x_1x_2}$ is the **correlation coefficient** between $\mathbf{x}_1$ and $\mathbf{x}_2$.

Similarly, the difference $\mathbf{w} = \mathbf{x}_1 - \mathbf{x}_2$ has a mean

$$\mu_w = \mu_{x_1} - \mu_{x_2}$$

and its variance is

$$\sigma_w^2 = \sigma_{x_1}^2 + \sigma_{x_2}^2 - 2\sigma_{x_1x_2} = \sigma_{x_1}^2 + \sigma_{x_2}^2 - 2\rho_{x_1x_2}\sigma_{x_1}\sigma_{x_2} \tag{C.7}$$

C.4 INDEPENDENCE AND UNCORRELATEDNESS

Two random variables **x** and **y** are said to be **independent** if their joint PDF, $f(x, y)$, can be separated; that is,

$$f(x, y) = f_1(x)f_2(y)$$

Two random variables **x** and **y** are said to be uncorrelated if their covariance vanishes as

$$\sigma_{xy} = E[(\mathbf{x} - \mu_x)(\mathbf{y} - \mu_y)] = 0$$

Since the correlation coefficient between **x** and **y** is defined as

$$\rho_{xy} = \frac{\sigma_{xy}}{\sigma_x\sigma_y}$$

we can also say that random variables **x** and **y** are **uncorrelated** if $\rho_{xy} = 0$.

When random variables **x** and **y** are independent, we find that

$$\sigma_{xy} = E[(\mathbf{x} - \mu_x)(\mathbf{y} - \mu_y)] = E[(\mathbf{x} - \mu_x)]E[(\mathbf{y} - \mu_y)] = 0$$

Thus, **x** and **y** are also uncorrelated. This proves that "independence implies uncorrelatedness." However, the converse statement ("lack of correlation implies independence") is NOT true, as the following example shows.

EXAMPLE

Let **z** be a random variable with a uniform PDF, $f(z) = 1/2$ for $z \in [-1, 1]$. Let the random variable $\mathbf{x} = \mathbf{z}$ and the random variable $\mathbf{y} = \mathbf{z}^2$.

It is obvious that **x** and **y** are not independent since $\mathbf{y} = \mathbf{x}^2$. From Appendix A, and the fact that **z** is uniformly distributed in $[-1, 1]$, we have

$$E[\mathbf{z}] = (a + b)/2 = (-1 + 1)/2 = 0 = E[\mathbf{x}] = \mu_x$$

Also, $E[\mathbf{y}] = E[\mathbf{z}^2] = \int_{-1}^{1}(1/2)z^2\,dz = 1/3 = \mu_y$. To compute σ_{xy}, the covariance of **x** and **y**, we obtain

$$E[(\mathbf{x} - \mu_x)(\mathbf{y} - \mu_y)] = E[\mathbf{x}(\mathbf{y} - 1/3)] = E[\mathbf{z}^3 - \mathbf{z}/3] = \int_{-1}^{1}(z^3 - z/3)/2\,dz = 0$$

Hence, **x** and **y** are uncorrelated.

C.5 PDF OF $\mathbf{z} = \mathbf{x} + \mathbf{y}$

Theorem C.5.1 *Suppose two random variables* **x** *and* **y** *are jointly normal with joint PDF*

$$f(x, y) = \frac{1}{2\pi\sqrt{1 - r^2}\sigma_1\sigma_2} \cdot e^{-\frac{1}{2(1-r^2)}\left[\frac{(x-\mu_1)^2}{\sigma_1^2} - \frac{2r(x-\mu_1)(y-\mu_2)}{\sigma_1\sigma_2} + \frac{(y-\mu_2)^2}{\sigma_2^2}\right]}$$

then the random variable $\mathbf{z} = \mathbf{x} + \mathbf{y}$ *is also normal and has a PDF*

$$g(z) = \frac{1}{\sqrt{2\pi(\sigma_1^2 + 2r\sigma_1\sigma_2 + \sigma_2^2)}} \cdot e^{-\frac{(z-\mu_1-\mu_2)^2}{2(\sigma_1^2+2r\sigma_1\sigma_2+\sigma_2^2)}}$$

Proof. We start with the definition of the CDF of **z**

$$G(z) = P(\mathbf{z} \leq z) = P(\mathbf{y} < \infty, \mathbf{x} \leq z - y) = \int_{-\infty}^{\infty}\int_{-\infty}^{z-y} f(x, y)\,dx\,dy$$

Now since the PDF is the derivative of the CDF, or $g(z) = dG(z)/dz$, we have

$$g(z) = \int_{-\infty}^{\infty} f(z-y, y)\,dy$$

$$= \frac{1}{2\pi\sqrt{1-r^2}\sigma_1\sigma_2}\int_{-\infty}^{\infty} e^{-\frac{1}{2(1-r^2)}\left[\frac{(z-y-\mu_1)^2}{\sigma_1^2} - \frac{2r(z-y-\mu_1)(y-\mu_2)}{\sigma_1\sigma_2} + \frac{(y-\mu_2)^2}{\sigma_2^2}\right]} dy$$

$$= \frac{1}{2\pi k}\int_{-\infty}^{\infty} e^{-\frac{1}{2k^2}[h(y,z)]}\,dy \quad .$$

where

$$k = \sqrt{1-r^2}\sigma_1\sigma_2$$

$$h(y, z) = (z-y-\mu_1)^2\sigma_2^2 - 2r\sigma_1\sigma_2(z-y-\mu_1)(y-\mu_2) + (y-\mu_2)^2\sigma_1^2$$

If we denote the half-length and the midpoint of the interval $[y-z-\mu_1, y-\mu_2]$ as Δ and u, respectively, then $\Delta = (z-\mu_1-\mu_2)/2$ and $u = y - z + \mu_1 + \Delta$. The expression $h(y, z)$ can be reduced to

$$h(y, z) = (u-\Delta)^2\sigma_2^2 + 2r\sigma_1\sigma_2(u-\Delta)(u+\Delta) + (u+\Delta)^2\sigma_1^2$$

$$= u^2(\sigma_1^2 + 2r\sigma_1\sigma_2 + \sigma_2^2) + 2u\Delta(\sigma_1^2 - \sigma_2^2) + \Delta^2(\sigma_1^2 - 2r\sigma_1\sigma_2 + \sigma_2^2)$$

$$= (k_1 u + k_2)^2 + \frac{4\Delta^2 k^2}{k_1^2}$$

$$k_1 = \sqrt{\sigma_1^2 + 2r\sigma_1\sigma_2 + \sigma_2^2}$$

$$k_2 = \Delta(\sigma_1^2 - \sigma_2^2)/k_1$$

Hence, $g(z)$ becomes

$$g(z) = \frac{1}{2\pi k}\int_{-\infty}^{\infty} e^{-\frac{1}{2k^2}\left[(k_1 u + k_2)^2 + \frac{4\Delta^2 k^2}{k_1^2}\right]} du$$

$$= \frac{1}{2\pi k} e^{-\frac{2\Delta^2}{k_1^2}} \int_{-\infty}^{\infty} e^{-\frac{1}{2}\left[\frac{k_1}{k}u + \frac{k_2}{k}\right]^2} \cdot \left(\frac{k_1}{k}du\right) \cdot \frac{k}{k_1}$$

With the change of variable (from u to v) via $v = (k_1/k)u + k_2/k$, we obtain the PDF for the random variable $\mathbf{z}$ as

$$g(z) = \frac{1}{2\pi k_1} e^{-\frac{2\Delta^2}{k_1^2}} \int_{-\infty}^{\infty} e^{-\frac{v^2}{2}}\,dv$$

$$= \frac{1}{2\pi k_1} e^{-\frac{2\Delta^2}{k_1^2}} \cdot \sqrt{2\pi}$$

$$= \frac{1}{\sqrt{2\pi}\,k_1} e^{-\frac{(z-\mu_1-\mu_2)^2}{2k_1^2}}$$

which is normally distributed such that $\mu_z = \mu_1 + \mu_2$ and $\sigma_z^2 = \sigma_1^2 + 2r\sigma_1\sigma_2 + \sigma_2^2$.

C.6 PDF OF $\mathbf{z} = \mathbf{x} - \mathbf{y}$

We can similarly prove the following theorem for the difference.

Theorem C.6.1 *Suppose two random variables* $\mathbf{x}$ *and* $\mathbf{y}$ *are jointly normal with joint PDF*

$$f(x, y) = \frac{1}{2\pi\sqrt{1 - r^2}\sigma_1\sigma_2} \cdot e^{-\frac{1}{2(1-r^2)}\left[\frac{(x-\mu_1)^2}{\sigma_1^2} - \frac{2r(x-\mu_1)(y-\mu_2)}{\sigma_1\sigma_2} + \frac{(y-\mu_2)^2}{\sigma_2^2}\right]}$$

then the random variable $\mathbf{z} = \mathbf{x} - \mathbf{y}$ *is also normal and has a PDF with mean* $(\mu_1 - \mu_2)$ *and variance* $(\sigma_1^2 + 2r\sigma_1\sigma_2 + \sigma_2^2)$.

C.7 PDF OF $\mathbf{z} = \mathbf{x}/\mathbf{y}$

Theorem C.7.1 *Suppose two random variables* $\mathbf{x}$ *and* $\mathbf{y}$ *are jointly normal and* $E[\mathbf{x}] = E[\mathbf{y}] = 0$, *with*

$$f(x, y) = \frac{1}{2\pi\sqrt{1 - r^2}\sigma_1\sigma_2} \cdot e^{-\frac{1}{2(1-r^2)}\left[\frac{x^2}{\sigma_1^2} - \frac{2rxy}{\sigma_1\sigma_2} + \frac{y^2}{\sigma_2^2}\right]}$$

then the random variable $\mathbf{z} = \mathbf{x}/\mathbf{y}$ *has a Cauchy PDF*

$$g(z) = \frac{\sigma_1\sigma_2\sqrt{1 - r^2}/\pi}{\sigma_2^2(z - r\sigma_1/\sigma_2)^2 + \sigma_1^2(1 - r^2)}$$

Proof. We start with the definition of the CDF of $\mathbf{z}$

$$G(z) = P(\mathbf{z} \le z) = P(\mathbf{y} < \infty, \mathbf{x} \le zy) = \int_{-\infty}^{\infty}\int_{-\infty}^{zy} f(x, y)\, dx\, dy$$

Now since the PDF is the derivative of CDF, or $g(z) = dG(z)/dz$, we have

$$
\begin{aligned}
g(z) &= \int_{-\infty}^{\infty} |y| f(zy, y)\, dy \\
&= \frac{1}{2\pi\sqrt{1-r^2}\sigma_1\sigma_2} \int_{-\infty}^{\infty} |y| e^{-\frac{1}{2\sigma_1^2\sigma_2^2(1-r^2)}[z^2y^2\sigma_2^2 - 2ry^2z\sigma_1\sigma_2 + y^2\sigma_1^2]}\, dy \\
&= \frac{1}{\pi\sqrt{1-r^2}\sigma_1\sigma_2} \int_{0}^{\infty} y e^{-\frac{y^2}{2\sigma_1^2\sigma_2^2(1-r^2)}[z^2\sigma_2^2 - 2rz\sigma_1\sigma_2 + \sigma_1^2]}\, dy \\
&= \frac{1}{\pi\sqrt{1-r^2}\sigma_1\sigma_2} \int_{0}^{\infty} y e^{-\frac{y^2}{2\sigma_1^2\sigma_2^2(1-r^2)}[(z\sigma_2 - r\sigma_1)^2 + \sigma_1^2(1-r^2)]}\, dy \\
&= \frac{\sigma_1\sigma_2\sqrt{1-r^2}/\pi}{\sigma_2^2(z - r\sigma_1/\sigma_2)^2 + \sigma_1^2(1-r^2)}
\end{aligned}
$$

D

GPS Dilution of Precisions

In the Global Positioning System (GPS), Dilution of Precision (DOP) factors are used to describe the effect of satellite geometry on the accuracy of the navigation solution.

The five DOPs (DOP in time or clock offset, DOP in vertical position, DOP in horizontal position, DOP in position, geometric DOP) are abbreviated as TDOP, VDOP, HDOP, PDOP, and GDOP, respectively. They are all ratios of standard deviations as defined in the following.

$$\text{TDOP} = \frac{c\sigma_{dt}}{\sigma_0}$$

$$\text{VDOP} = \frac{\sigma_U}{\sigma_0}$$

$$\text{HDOP} = \frac{\text{DRMS}_{2D}}{\sigma_0} = \frac{\sqrt{\sigma_E^2 + \sigma_N^2}}{\sigma_0}$$

$$\text{PDOP} = \frac{\text{DRMS}_{3D}}{\sigma_0} = \frac{\sqrt{\sigma_E^2 + \sigma_N^2 + \sigma_U^2}}{\sigma_0}$$

$$\text{GDOP} = \frac{\sqrt{c^2\sigma_{dt}^2 + \sigma_E^2 + \sigma_N^2 + \sigma_U^2}}{\sigma_0}$$

where c represents the speed of light and σ_0 denotes the standard deviation of the observed pseudo ranges.The term σ_{dt} is the standard deviation of time errors, and $\sigma_E, \sigma_N, \sigma_U$ are the standard deviations of the one-dimensional errors in the East, North, and Up directions, respectively.

We show an example to compare the specifications of two GPSs.

EXAMPLE D.1

If system A has a quoted accuracy of 3 meters DRMS_{3D} and system B has a quoted CEP of 2 meters, which system is more accurate?

The answer depends on the ratio of PDOP to HDOP. Suppose that PDOP/ HDOP $= 2.1$, and let $x = \text{DRMS}_{3D}/\text{CEP}$

```
m=10;
dx=(b-a)/m;

  for k=1:m-1
     x=a+k*dx;
     y=nf1(x);
     plot([x,x],[0,y],':')
  end
  ya=nf1(a);
  yb=nf1(b);
  plot([a,a],[0,ya],'--')
  plot([b,b],[0,yb],'--')

% NEED TO INCLUDE STYLED TEXT TOOLBOX IN MATLAB PATH
sxlabel('{\i x}')
sylabel('{\i f(x)} ')
t=['Area under Normal PDF {\i f(x)}, for {\i x} in [' num2str(-r) ',',...
  num2str(r) '] is ' num2str(p,4) ];
stitle(t)
figure(1)
pf3('f11')  % or pf3('f33')
---------------------------------------
```

```
nf1.m
---------------------------------------
function y=nf1(x)
% nf1.m  3-06-96
%
% Spatial Error Analysis ToolBox Version 1.0, October 5, 1997
% Copyright 1997-1998 by David Y. Hsu  All rights reserved.
% dhsu@littongcs.com
%
% define f(x) = N(0,1) = Standard Normal Probability Density Function
% notice .^  is used  for vector x, sigma=1
fac1=1/sqrt(2*pi);
y=fac1*exp(-x.^2/2);
---------------------------------------
```

```
nf2.m
---------------------------------------
function p=nf2(r)
% nf2.m  3-06-96
%
% Spatial Error Analysis ToolBox Version 1.0,  October 5, 1997
% Copyright 1997-1998 by David Y. Hsu  All rights reserved.
% dhsu@littongcs.com
%
```

```
% given r>=0, find p=F(r)
% F(r) = integral of N(0,1) over [-r ,r]. (using erf(x) relation)

if r<0
 error('check input argument in nf2(r) ')
end

p=erf(r/sqrt(2));
----------------------------------------

pf3.m
----------------------------------------
function pf3(fname)
% pf3.m  01-25-96
% print figure to an eps file
%
% Spatial Error Analysis ToolBox Version 1.0, October 5, 1997
% Copyright 1997-1998 by David Y. Hsu All rights reserved.
% dhsu@littongcs.com
%
fname=['c:\bookeps\' fname '.eps'];
s1=['printsto -deps ' fname];
eval(s1)
----------------------------------------

newton.m
-------------------------------------------
function x=newton(xo,fun,dfun,tol)
% newton.m  1-1-97
%
% Spatial Error Analysis ToolBox Version 1.0, October 5, 1997
% Copyright 1997-1998 by David Y. Hsu All rights reserved.
% dhsu@littongcs.com
%
% xo is initial guess
% fun and dfun are string names

% example:
%        fun1 = sf3a;
%       dfun1 = sf1 ;
%     x = newton(xo,'sf3a','sf1',0.0001);
global p

 eval(['x = xo - ', fun,'(xo)/', dfun,'(xo);']);
 d=abs(x-xo);
 N=1;
 while d>tol
   xo=x ;
```

```
  eval(['x = xo - ', fun,'(xo)/', dfun,'(xo);']);
  N=N+1;
  d=abs(x-xo);
 end
--------------------------------------------
```

E.2 M-FILES USED IN CHAPTER 2

```
rmsx.m
----------------------------------------
function r = rmsx(x)
% rmsx.m 8-06-96
%
% Spatial Error Analysis ToolBox Version 1.0, October 5, 1997
% Copyright 1997-1998 by David Y. Hsu All rights reserved.
% dhsu@littongcs.com
%
% find  sqrt{[x(1)^2+x(2)^2+ ... + x(N)^2]/N}
N=length(x);
r =sqrt(sum(x.^2)/N);
----------------------------------------
```

```
nf1.m (see M-files used in Chapter 1)
----------------------------------------
```

```
tandn.m
----------------------------------------
% tandn.m   3-23-96   plot Figure 2.3
%
% Spatial Error Analysis ToolBox Version 1.0,  October 5, 1997
% Copyright 1997-1998 by David Y. Hsu All rights reserved.
% dhsu@littongcs.com
%
clear global n
global n
n=100;
subplot(211)
fplot('nf1',[-5,5])
sxlabel('{\i x}')
sylabel('{\i f(x)}')
stitle('{\i N(0,1)} PDF')
axis([-5 , 5, 0, 0.5])
grid

n=100;
```

```
subplot(212)
fplot('tdis',[-5,5])
grid
axis([-5 , 5, 0, 0.5])
sxlabel('{\i \tau}')
sylabel('{\i g(\tau{})}')
stitle('Student-{\i {t_n} } PDF, {\i n} = 100')
fixstext
figure(1)
pf3('f23')
---------------------------------------
```

xandn.m

```
---------------------------------------
% xandn.m   9-26-97   plot Figure 2.6
%
% Spatial Error Analysis ToolBox Version 1.0, October 5, 1997
% Copyright 1997-1998 by David Y. Hsu All rights reserved.
% dhsu@littongcs.com
%
clear global n
global n
clf

subplot(211)
fplot('nf1',[-5,5])
sylabel('{\i f(x)}')
sxlabel('{\i x}')
stitle(['{\i f(x) = N(0,1)} is defined for all real {\i u}'])
axis([-5 , 5, 0, 0.5])
grid

n=100;
umin = -sqrt(2*n-1)
subplot(212)
fplot('gofu',[-5,5])
sylabel('{\i g(u)}')
sxlabel('{\i u}')
stitle(['{\i g(u)} is defined for {\i u \geq } ' num2str(umin), ', when
    {\i n} = 100'])
axis([-5,5, 0, 0.5])
grid
figure(1)
pf3('f26')
---------------------------------------
```

tdis.m

```
----------------------------------------
function y=tdis(x)
% tdis.m   3-22-96
%
% Spatial Error Analysis ToolBox Version 1.0, October 5, 1997
% Copyright 1997-1998 by David Y. Hsu All rights reserved.
% dhsu@littongcs.com
%
% student t - Probability Density Function
% notice .* and .^ are used for vector x:
global n
fac1=gamma((n+1)/2)/gamma(n/2)/sqrt(pi*n) ;
y=fac1*(1+x.^2/n).^(-(n+1)/2);
----------------------------------------
```

gofu.m

```
----------------------------------------
% gofu.m  09-26-97
% g(u), Probability Density Function of the random variable u
%
% Spatial Error Analysis ToolBox Version 1.0, October 5, 1997
% Copyright 1997-1998 by David Y. Hsu All rights reserved.
% dhsu@littongcs.com
%
function y=gofu(u)
global n
xm=sqrt(2*n-1);
fac1=1/(2^(n-1)*gamma(n/2));
y=fac1*(u+xm).^(n-1).*exp(-(u+xm).^2/4);
----------------------------------------
```

x2df.m

```
------------------------------------------
function y=x2df(x)
% x2df.m 12-30-96
%
% Spatial Error Analysis ToolBox Version 1.0, October 5, 1997
% Copyright 1997-1998 by David Y. Hsu All rights reserved.
% dhsu@littongcs.com
%
global n
if n<2
  fac1=1/(2^(n/2+1)*gamma(n/2+1));
  y=fac1*x.^(n/2).*exp(-x/2);
else
  fac1=1/(2^(n/2)*gamma(n/2));
  y=fac1*x.^(n/2-1).*exp(-x/2);
end
```

```
----------------------------------------
x2dg.m
----------------------------------------
function y=x2dg(x,n)
% x2dn.m 1-2-96
%
% Spatial Error Analysis ToolBox Version 1.0, October 5, 1997
% Copyright 1997-1998 by David Y. Hsu All rights reserved.
% dhsu@littongcs.com
%
if n<2
fac1=1/(2^(n/2)*gamma(n/2+1));
y=fac1*x.^(n/2).*exp(-x/2);
else
 y=0;
end
----------------------------------------

x2cdf.m
----------------------------------------
function p=x2cdf(x)
% x2cdf.m 1-16-96
%
% Spatial Error Analysis ToolBox Version 1.0, October 5, 1997
% Copyright 1997-1998 by David Y. Hsu All rights reserved.
% dhsu@littongcs.com
%
% where F(x,n) = [integral of f(x,n) from 0 to a]
%   f(x,n) is the chi-square PDF with degree n
%
% when integrating f(t,n) (implemented in x2df.m) from 0 to x, for n<2
% use the F(x,n)=g(x,n)+F(x,n+2) relationship, where
% g(x,n)= (x/2)^(n/2).*exp(-x/2)/gamma(n/2+1) -- implemented in x2dg.m
%
global n
if n<2
p=quad8('x2df',0,x) + x2dg(x,n);
else
p= quad8('x2df',0,x);
end
----------------------------------------

cnf.m
----------------------------------------
function p=cnf(ep,n)
% cnf.m 7-01-94
% determine confidence coeff about s, the sample estimate of population
% sigma for a given sample size and variation epsilon (in % of s)
%
```

```
% Spatial Error Analysis ToolBox Version 1.0, October 5, 1997
% Copyright 1997-1998 by David Y. Hsu All rights reserved.
% dhsu@littongcs.com
%
%          n       _
%    s^2 =SUM(xi - x)^2/(n-1)  is the sample estimate for sigma^2
%         i=1
% without the next line, n will stay the same after the first use.
clear global n

% n is used in the definition of chi-square CDF: x2cdf below.
global n
n=n-1;
if n<0 | ep<0
disp(' *** error in input *** ')
else
  if n<101 & n>=0              %  small sample
    if n==0
     p=0;
    else    %     s(1-epsilon) < sigma < s(1+epsilon)
     a=n/(1+ep)^2 ;
     b=n/(1-ep)^2 ;
     p=x2cdf(b)-x2cdf(a);
    end
  else                         %  large sample
     a=sqrt(2*n)/(1+ep)-sqrt(2*n-1);
     b=sqrt(2*n)/(1-ep)-sqrt(2*n-1);
     p=quad8('nf1',a,b);
  end
end
----------------------------------------

mucnf.m
----------------------------------------
function p=mucnf(ep,n)
% mucnf.m 3-07-96
% determine confidence level about xb, the sample estimate of population
% mean mu, for a given sample size and variation epsilon (in % of s).
%
% Spatial Error Analysis ToolBox Version 1.0, October 5, 1997
% Copyright 1997-1998 by David Y. Hsu All rights reserved.
% dhsu@littongcs.com
%
% without the next line, n will stay the same after the first use.
clear global n
% n is used in the definition of student-t distribution below.
global n
capn=n;
```

```
n=n-1;
if n<0 | ep<0
disp(' *** error in input *** ')
else
  b=sqrt(capn)*ep ;

  if n<101 & n>=0                % small sample
     if n==0
      p=0;
     else
%              xb - epsilon * s  < mu < xb + epsilon * s
      p=2*quad('tdis',0,b);
     end
  else                         % large sample
     if b>4
      p=1;
     else
      p=2*quad('nf1',0,b);
     end
  end
end
-----------------------------------------
```

E.3 M-FILES USED IN CHAPTER 3

```
nf1.m (see M-files used in Chapter 1)
-------------------------------------------

nf2.m (see M-files used in Chapter 1)
-------------------------------------------

nf3.m
-------------------------------------------
function r=nf3(p)
% nf3.m  3-07-96
%
% Spatial Error Analysis ToolBox Version 1.0, October 5, 1997
% Copyright 1997-1998 by David Y. Hsu All rights reserved.
% dhsu@littongcs.com
%
% given p and f(x), find r such that G(r) = F(r)-p = 0,
% where F(r) = integral of f(x) from -r to r,
% using inverse erf(x) relation.
%  f(x)= N(0,1) used here.
%
```

```
if p>1 | p<0
 disp(' input error for nf3(p)')
 disp(' 0 <= p <= 1')
else
   if p==1
     r=Inf;
   else
     r=sqrt(2)*erfinv(p);
   end
end
-----------------------------------------
```

```
nf6.m
-----------------------------------------
function y=nf6(x,mu,sig)
% nf6.m 3-06-96
%
% Spatial Error Analysis ToolBox Version 1.0,  October 5, 1997
% Copyright 1997-1998 by David Y. Hsu  All rights reserved.
% dhsu@littongcs.com
%
% nf6.m  define f(x) = N(mean, variance)
% notice .^  is used  for vector x
fac1=1/(sqrt(2*pi)*sig);
y=fac1*exp(-(x-mu).^2/(2*sig^2));
-----------------------------------------
```

```
nf7.m
-----------------------------------------
%nf7.m
% nf7.m  3-06-96
%
% Spatial Error Analysis ToolBox Version 1.0, October 5, 1997
% Copyright 1997-1998 by David Y. Hsu  All rights reserved.
% dhsu@littongcs.com
%
function nf7(p)
clf

 load c:\mfile\eg3p2.mat

mu=mean(z)
sig=std(z)

  xx1=floor(mu-4*sig);
  xx2=ceil(mu+4*sig);
  d=(xx2-xx1)/100;
  x=[xx1:d:xx2];
```

```
    y=nf6(x,mu,sig);
    plot(x,y)
    hold

  r=nf3(p)
  a=sig*(-r)+mu;
  b=sig*(r)+mu;

  m=10;
  dx=(b-a)/m;

    for k=1:m-1
       x=a+k*dx;
       y=nf6(x,mu,sig);
       plot([x,x],[0,y],':')
    end
    ya=nf6(a,mu,sig);
    yb=nf6(b,mu,sig);
    plot([a,a],[0,ya],'--')
    plot([b,b],[0,yb],'--')


  sxlabel('{\i z} ')
  sylabel('{\i g(z)} ')
  t=['Area under Normal PDF {\i g(z)}, for {\i z} in [' num2str(a) ','
       num2str(b) '] is ' num2str(p) ];
  stitle(t)
  figure(1)
  pf3('f34')
  ------------------------------------------

  r2p1d.m
  ------------------------------------------
  function r2p1d(k)
  % r2p1d.m 5-04-96
  %
  % Spatial Error Analysis ToolBox Version 1.0, October 5, 1997
  % Copyright 1997-1998 by David Y. Hsu All rights reserved.
  % dhsu@littongcs.com
  %
  % given k = multiple of SIGMA (or RMS)
  % given k = multiple of MAE
  % given k = multiple of LEP
  % r = k * SIGMA
  % r = k * MAE        (MAE = Mean Absolute Error)
  % r = k * LEP        (LEP = Linear Error Probable)
  % find p = F(r) = integral of f(x) from -r to r
  % f(x) is 1-dim normal distribution N(0,1)
```

```
disp( ' ')
disp( ' ')
ic=menu('Options:','k*SIGMA','k*MAE','k*LEP','all the above');

if ic==1
r= k;
p=nf2(r) ;
disp(['For r=' ,num2str(k), ' * SIGMA'])
disp(['p=', num2str(p)])
end

if ic==2
r= k* sqrt(2/pi);
p=nf2(r)    ;
disp(['For r=' ,num2str(k), ' * MAE'])
disp(['p=', num2str(p)])
end

if ic==3
r0=nf3(0.5);
r=k*r0;
p=nf2(r) ;
disp(['For r=' ,num2str(k), ' * LEP'])
disp(['p=', num2str(p)])
end

if ic==4
r= k;
p=nf2(r)     ;
disp(['For r=' ,num2str(k), ' * SIGMA'])
disp(['   p=', num2str(p)])

r= k* sqrt(2/pi);
p=nf2(r)      ;
disp(['For r=' ,num2str(k), ' * MAE'])
disp(['   p=', num2str(p)])

r=k*nf3(0.5);
p=nf2(r)      ;
disp(['For r=' ,num2str(k), ' * LEP'])
disp(['   p=', num2str(p)])
end
------------------------------------------
p2r1d.m
------------------------------------------
function p2r1d(p)
% p2r1d.m 5-04-96
%
```

```
% Spatial Error Analysis ToolBox Version 1.0, October 5, 1997
% Copyright 1997-1998 by David Y. Hsu All rights reserved.
% dhsu@littongcs.com
%
% 1-dim normal distribution
% given p and f(x), find r such that F(r) - p = 0,
% express r in terms of SIGMA, MAE, LEP
% where F(r) = integral of f(x) from -r to r.
%  standard noraml distribution N(0,1) used for f(x) here.
%
r=nf3(p);

disp('  ')
disp('  ')
disp([' For p=', num2str(p) ])
ic=menu('options:', 'k*SIGMA', 'k*MAE', 'k*LEP', 'all the above');

if ic==1
disp(['r=', num2str(r), ' * SIGMA'])
end

if ic==2
r=r*sqrt(pi/2);
disp(['r=', num2str(r), ' * MAE'])
end

if ic==3
r=r/nf3(0.5);
disp(['r=', num2str(r), ' * LEP'])
end

if ic==4
disp(['r=', num2str(r), ' * SIGMA'])

r=r*sqrt(pi/2);
disp(['r=', num2str(r), ' * MAE'])

r=nf3(p)/nf3(0.5);
disp(['r=', num2str(r), ' * LEP'])
end
--------------------------------------------
```

E.4 M-FILES USED IN CHAPTER 4

```
nf2.m  (see M-files used in Chapter 1)
----------------------------------------

cf1.m
----------------------------------------
function y=cf1(r)
% cf1.m 6/04/97,  define f(r)= circular normal pdf
%
% Spatial Error Analysis ToolBox Version 1.0, October 5, 1997
% Copyright 1997-1998 by David Y. Hsu All rights reserved.
% dhsu@littongcs.com
%
% define f(r) = circular normal pdf
% Rayleigh Probability Density Function  sigma = 1
y=r.* exp(-r.^2/2);
----------------------------------------

cf2.m
----------------------------------------
function p=cf2(r)
% cf2.m  1-23-96
%
% Spatial Error Analysis ToolBox Version 1.0, October 5, 1997
% Copyright 1997-1998 by David Y. Hsu All rights reserved.
% dhsu@littongcs.com
%
% for 2-dim normal distribution with sigma(x) = sigma(y)
% find p = F(r)
% where F(r) = integral of f(x) from 0 to r.
% Rayleigh pdf used for f(x) here.
%

if r<0
  error('check input argument in cf2(r)')
end

p=1-exp(-r.^2/2);
----------------------------------------

cf3.m
----------------------------------------
function r=cf3(p)
% cf3.m  3-06-96
%
```

```
% Spatial Error Analysis ToolBox Version 1.0, October 5, 1997
% Copyright 1997-1998 by David Y. Hsu All rights reserved.
% dhsu@littongcs.com
%
% given p and f(x), find r such that G(r) = F(r) - p = 0,
% where F(r) = integral of f(x) from 0 to r.
%    Rayleigh pdf used for f(x) here.
%
if p>1 | p<0
 disp( 'input error for cf3(p)')
 disp( '0 <= p <= 1')
else
 if p==1
  r=Inf;
 else
  r=sqrt(-2*log(1-p));
 end
end
----------------------------------------

gf2.m
----------------------------------------
function p=gf2(r,mu)
% gf2.m 01-23-96
%
% Spatial Error Analysis ToolBox Version 1.0, October 5, 1997
% Copyright 1997-1998 by David Y. Hsu All rights reserved.
% dhsu@littongcs.com
%
% mu = min(sx,sy)/max(sx,sy)
% r = R/max(sx,sy)
%

% r >=0, mu in (0,1)
 if r<0 | mu<=0 | mu>=1
  error(' check input arguments in gf2(r,mu)')
 end

  n=2000;
  h=2*pi/n;
  fac1=1/(2*pi*mu);
  x=[0:n]*h;
  gf=(cos(x).^2+sin(x).^2/mu^2);
  y=(1-exp(-r^2.*gf/2))./gf;
  p=fac1*simprule(y,h);
----------------------------------------
```

```
simprule.m
---------------------------------------
function s=simprule(y,h)
% simprule.m 1-23-96
%
% Spatial Error Analysis ToolBox Version 1.0, October 5, 1997
% Copyright 1997-1998 by David Y. Hsu All rights reserved.
% dhsu@littongcs.com
%
n= length(y)-1   ;
s0=y(1)+y(n+1);
s1=sum(y(2:2:n));
s2=sum(y(3:2:n-1));
s=h*(s0+4*s1+2*s2)/3;
---------------------------------------
```

```
r2p2d.m
---------------------------------------
function r2p2d(k)
% r2p2d.m  6-07-96
%
% Spatial Error Analysis ToolBox Version 1.0, October 5, 1997
% Copyright 1997-1998 by David Y. Hsu All rights reserved.
% dhsu@littongcs.com
%
% for 2-dim normal distribution with sigma(x) = sigma(y), rho = 0.
% given k = multiple of SIGMA
% given k = multiple of MRE
% given k = multiple of DRMS(RMSR)
% given k = multiple of CEP
% r = k * SIGMA
% r = k * MRE          (MRE = Mean Radial Error)
% r = k * DRMS(RMSR)  (RMSR = Root Mean Square Radial Error)
%                      DRMS = Distance Root Mean Square
% r = k * CEP          (CEP = Circular Error Probable)
% find p = F(r)
% where F(x) = integral of f(x) from 0 to r.
% Rayleigh pdf used for f(x) here.
%

disp(' ')
disp(' ')
ic=menu('Options:','k*SIGMA','k*MRE','k*DRMS','k*CEP','all the
     above');

if ic==1
r= k;
p=cf2(r) ;
```

```
disp(['For r=' ,num2str(k), ' * SIGMA'])
disp(['p=', num2str(p)])
end

if ic==2
r= k* sqrt(pi/2);
p=cf2(r)   ;
disp(['For r=' ,num2str(k), ' * MRE'])
disp(['p=', num2str(p)])
end

if ic==3
r= k* sqrt(2);
p=cf2(r)   ;
disp(['For r=' ,num2str(k), ' * DRMS(RMSR)'])
disp(['p=', num2str(p)])
end

if ic==4
r=k*cf3(0.5);
p=cf2(r) ;
disp(['For r=' ,num2str(k), ' * CEP'])
disp(['p=', num2str(p)])
end

if ic==5
r= k;
p=cf2(r)    ;
disp(['For r=' ,num2str(k), ' * SIGMA'])
disp(['   p=', num2str(p)])

r= k* sqrt(pi/2);
p=cf2(r)     ;
disp(['For r=' ,num2str(k), ' * MRE'])
disp(['   p=', num2str(p)])

r= k* sqrt(2);
p=cf2(r)   ;
disp(['For r=' ,num2str(k), ' * DRMS(RMSR)'])
disp(['   p=', num2str(p)])

r=k*cf3(0.5);
p=cf2(r)    ;
disp(['For r=' ,num2str(k), ' * CEP'])
disp(['   p=', num2str(p)])
end
----------------------------------------
```

p2r2d.m

```
---------------------------------------
function p2r2d(p)
% p2r2d.m 6-07-96
%
% Spatial Error Analysis ToolBox Version 1.0, October 5, 1997
% Copyright 1997-1998 by David Y. Hsu All rights reserved.
% dhsu@littongcs.com
%
% given p and f(x), find r such that F(r) - p = 0,
% express r in terms of sigma, CEP, MRE, DRMS(RMSR)
% where F(r) = integral of f(x) from 0 to r.
%     Rayleigh pdf used for f(x) here.
%

disp('  ')
disp('  ')
disp([' For p=', num2str(p) ])
ic=menu('options:', 'k*sigma', 'k*MRE',...
            'k*CEP', 'k*DRMS(RMSR)', 'all 4 above');
r0=cf3(p) ;
%

if ic==1
disp(['r=', num2str(r0), ' * SIGMA'])
end

if ic==2
r1=r0/sqrt(pi/2);
disp(['r=', num2str(r1), ' * MRE'])
end

if ic==3
r2=r0/cf3(.5);
disp(['r=', num2str(r2), ' * CEP'])
end

if ic==4
r3=r0/sqrt(2);
disp(['r=', num2str(r3), ' * DRMS(RMSR)'])
end

if ic==5
disp(['r=', num2str(r0), ' * SIGMA'])

r1=r0/sqrt(pi/2);
disp(['r=', num2str(r1), ' * MRE'])
```

```
r2=r0/sqrt(2);
disp(['r=', num2str(r2), ' * DRMS(RMSR)'])

r3=r0/cf3(0.5);
disp(['r=', num2str(r3), ' * CEP'])
end
--------------------------------------
```

```
edraud.m
--------------------------------------
function p=edraud(a,b,sx,sy)
% edraud.m  2-07-96
%
% Spatial Error Analysis ToolBox Version 1.0, October 5, 1997
% Copyright 1997-1998 by David Y. Hsu  All rights reserved.
% dhsu@littongcs.com
%
% The probability of a
% 2D joint elliptical normal pdf with sx, sy, rho = 0
% over a rectangular region:  |x| < a,   |y| < b.
%

if sx<0 | sy<0 | a<0 | b<0 | (sx==0 & sy==0)
  error('check signs of input argument')
end

if sx==0
 p=nf2(b/sy);
end

if sy==0
 p=nf2(a/sx);
end

if sx > 0 & sy > 0
 p=nf2(a/sx)*nf2(b/sy);
end
--------------------------------------
```

```
edraui.m
--------------------------------------
function [a,b]=edraui(p,v,sx,sy)
% edraui.m  2-07-96
%
% Spatial Error Analysis ToolBox Version 1.0, October 5, 1997
% Copyright 1997-1998 by David Y. Hsu  All rights reserved.
% dhsu@littongcs.com
%
```

```
%    Rayleigh pdf used for f(x) here.
%
if p>1 | p<0 | sig <=0
 error( 'check input arguments in cdcaui.m')
end

  if p==1
   R=Inf;
  else
   R=sig*sqrt(-2*log(1-p));
  end
---------------------------------------
```

edcaud.m

```
---------------------------------------
function R=edcaui(p,a,b)
% edcaui.m  1-23-96
%
% Spatial Error Analysis ToolBox Version 1.0, October 5, 1997
% Copyright 1997-1998 by David Y. Hsu All rights reserved.
% dhsu@littongcs.com
%
% Elliptical PDF, Uncorrelated, Circular region
% function R = edcaui(p,a,b)

if p<0 | p>1 | a< 0 | b< 0  | (a==0 & b==0)
   error('check signs of inputs for edcaui(p,a,b) ')
end

R=1;
p0=edcaud(R,a,b);
%%%%%%%%%%%%%%%%%%%%%%%
for j=1:2:5
  while (p0 > p)
   R=R-(0.1)^j;
   p0=edcaud(R,a,b);
  end

  while (p0 < p)
   R=R+(0.1)^j;
   p0=edcaud(R,a,b);
  end
 end
---------------------------------------
```

edcaui.m

```
---------------------------------------
function p=edcaud(R,sx,sy)
```

```
% edcaud.m 1-23-96
%
% Spatial Error Analysis ToolBox Version 1.0, October 5, 1997
% Copyright 1997-1998 by David Y. Hsu All rights reserved.
% dhsu@littongcs.com
%
% Elliptical PDF, Uncorrelated, Circular region
% edcaud.m, Find probability  p, given R,sx,sy.
% bf1b.m is the normalized version
% function p = edcaud(R,sx,sy)
%   1>= mu >=0

if R<0 | sx< 0 | sy< 0 | (sx==0 & sy==0)
   error('check input  arguments for edcaud(R,sx,sy) ')
end

mu=min(sx,sy)/max(sx,sy);
r=R/max(sx,sy);
p=edca(r,mu);
----------------------------------------

edcacd.m
----------------------------------------
function p=edcacd(R,sx,sy,rho)
% edcacd.m 1-23-96
%
% Spatial Error Analysis ToolBox Version 1.0, October 5, 1997
% Copyright 1997-1998 by David Y. Hsu All rights reserved.
% dhsu@littongcs.com
%
% Elliptical PDF, correlated, Circular region

if R<0 | sx <0 | sy<0 | abs(rho)>1 | (sx==0 & sy==0)
  error('check input arguments in edcacd.m')
end

if abs(rho)==1
  st=sqrt(sx^2+sy^2);
  p=nf2(R/st);
end

if abs(rho)==0
  p=edcaud(R,sx,sy);
end

if abs(rho)<1 & abs(rho) >0
  A= 1/(1-rho^2)/sx^2  ;
  B= -2*rho/(1-rho^2)/(sx*sy) ;
```

```
  C= 1/(1-rho^2)/sy^2 ;
  [a1, c1]=abc2ac(A,B,C);
  s1=1/sqrt(a1);
  s2=1/sqrt(c1);
  p=edcaud(R,s1,s2);
end
----------------------------------------

edcaci.m
----------------------------------------
function R=edcaci(p,sx,sy,rho)
% edcaci.m 1-23-96
%
% Spatial Error Analysis ToolBox Version 1.0, October 5, 1997
% Copyright 1997-1998 by David Y. Hsu All rights reserved.
% dhsu@littongcs.com
%
% Elliptical PDF, correlated, Circular region

if p<0 | p>1 | sx <0 | sy<0 | abs(rho)>1 | (sx==0 & sy==0)
   error('check input arguments in edcacd.m')
end


R=1;
p0=edcacd(R,sx,sy,rho)
%%%%%%%%%%%%%%%%%%%%%%
for j=1:2:5
  while (p0 > p)
    R=R-(0.1)^j;
    p0=edcacd(R,sx,sy,rho)
  end

  while (p0 < p)
    R=R+(0.1)^j;
    p0=edcacd(R,sx,sy,rho)
  end
end
----------------------------------------

edeaud.m
----------------------------------------
function p=edeaud(a,b,sx,sy)
% edeaud.m 1-23-96
%
% Spatial Error Analysis ToolBox Version 1.0, October 5, 1997
% Copyright 1997-1998 by David Y. Hsu All rights reserved.
% dhsu@littongcs.com
```

```
%
% edea = elliptical normal pdf over elliptical area,
% elliptical normal pdf, x and y are uncorrelated.
% 1/(2*pi*sx*sy) exp[(-x^2/sx^2 - y^2/sy^2)/2]
% elliptical area of integration
%   x^2/a^2 + y^2/b^2 = 1
%
% example
% a=0.6, b=1.2, sx=.4, sy=1.0.
% p=edeaud(0.6, 1.2, 0.4, 1.0)
%

if sx< 0 | sy< 0 | a<=0 | b<=0 | (sx==0 & sy==0)
 error('check input arguments in edeaud.m')
end

s1=sx/a;
s2=sy/b;
mu=min(s1,s2)/max(s1,s2) ;
p=edcaud(1,s1,s2);
----------------------------------------

edeaui.m
----------------------------------------
function [a,b]=edeaui(p,v,sx,sy)
% edeaui.m 1-23-96
%
% Spatial Error Analysis ToolBox Version 1.0, October 5, 1997
% Copyright 1997-1998 by David Y. Hsu All rights reserved.
% dhsu@littongcs.com
%
% elliptical normal pdf, x and y are uncorrelated
% 1/(2*pi*sx*sy) exp[(-x^2/sx^2 - y^2/sy^2)/2]
% elliptical area of integration
%   x^2/a^2 + y^2/b^2 = 1
%
% v = a/b
%

if sx< 0 | sy< 0 | v <0 | (sx==0 & sy==0)
 error('check input arguments in edeaui.m')
end

b=1;
a=v*b;
p0=edeaud(a,b,sx,sy);

%%%%%%%% searching for a, b %%%%%%%%%%%%%%%
```

```
for j=1:2:5
 while (p0 > p)
  b=b-(0.1)^j;
  a=v*b;
  p0=edeaud(a,b,sx,sy);
 end

 while (p0 < p)
  b=b+(0.1)^j;
  a=v*b;
  p0=edeaud(a,b,sx,sy);
 end
end
----------------------------------------
```

edeacd.m

```
----------------------------------------
function p=edeacd(a,b,sx,sy,rho)
% edeacd.m  1-23-96
%
% Spatial Error Analysis ToolBox Version 1.0, October 5, 1997
% Copyright 1997-1998 by David Y. Hsu  All rights reserved.
% dhsu@littongcs.com
%
% function p = edeacd(a,b,sx,sy,rho)
% edeac = elliptical normal pdf over elliptical area, x,y are correlated
% elliptical normal pdf
% 1/[2*pi*sx*sy*sqrt(1-rho^2)]*
%   exp{-[x^2/sx^2-2*rho*x*y/(sx*sy)+y^2/sy^2]/(2(1-rho^2)}
% elliptical area of integration
%   x^2/a^2 + y^2/b^2 = 1
%
% example
% a = 0.6; b = 1.2; sx = 0.4; sy = 1; rho = 0.3
% p = edeacd(0.6, 1.2, 0.4, 1, 0.3)
%

 if abs(rho)>1 | sx<=0 | sy<=0 | a<=0 | b<=0
   error('check input arguments in edeacd(a,b,sx,sy,rho)')
 end

   if abs(rho)==1
    u=sy/sx;
    st=sqrt((1+u^2))*sx;
    t0=sqrt(1+u^2)*a*b/sqrt(a^2*u^2+b^2);
    p=nf2(t0/st);
   end
```

```
  if rho==0
    p=edeaud(a,b,sx,sy)    ;
  end

  if abs(rho)<1 & abs(rho) >0
   sig1=sx/a;
   sig2=sy/b;

   A=1/sig1^2;
   B=-2*rho/(sig1*sig2);
   C=1/sig2^2;

   [a1, c1]=abc2ac(A,B,C);

   s1=1/sqrt(a1);
   s2=1/sqrt(c1);

   f0=sqrt(1-rho^2);

   t1=s1*f0;
   t2=s2*f0;

   p=edcaud(1,t1,t2) ;
  end
----------------------------------------
edeaci.m
----------------------------------------
function [a,b]=edeaci(p,v,sx,sy,rho)
% edeaci.m  1-23-96
%
% Spatial Error Analysis ToolBox Version 1.0, October 5, 1997
% Copyright 1997-1998 by David Y. Hsu  All rights reserved.
% dhsu@littongcs.com
%
% function [a,b] = edeaci(p,v,sx,sy,rho)
% edeac = elliptical normal pdf over elliptical area, x,y are correlated
% elliptical normal pdf
% 1/[2*pi*sx*sy*sqrt(1-rho^2)]
%   exp{-[x^2/sx^2-2*rho*x*y/(sx*sy)+y^2/sy^2]/(2(1-rho^2)}
% elliptical area of integration
%   x^2/a^2 + y^2/b^2 = 1
%
%        v=a/b
%
% example
% p = 0.7; v = 0.5; sx = 0.4; sy = 1; rho = 0.3
% [a,b] = edeaci(0.7, 0.5, 0.4, 1, 0.3)
%
```

```
if abs(rho)>1 | v<=0 |  sx<=0 | sy<=0
  error('check input arguments in edeacd.m ')
end


b=1;
a=v*b;
p0=edeacd(a,b,sx,sy,rho);

%%%%%%%% searching for a, b %%%%%%%%%%%%%%%
for j=1:2:5
 while (p0 > p)
   b=b-(0.1)^j;
   a=v*b;
   p0=edeacd(a,b,sx,sy,rho);
 end

 while (p0 < p)
   b=b+(0.1)^j;
   a=v*b;
   p0=edeacd(a,b,sx,sy,rho);
 end
end
----------------------------------------

abc2ac.m
----------------------------------------
function [a1,c1]=abc2ac(a,b,c)
% abc2ac.m  2-04-96
%
% Spatial Error Analysis ToolBox Version 1.0, October 5, 1997
% Copyright 1997-1998 by David Y. Hsu  All rights reserved.
% dhsu@littongcs.com
%
% rotation of ellipse

ang=atan2(b,a-c);
t=ang/2;
tdeg=t*180/pi;

ct=cos(t);
st=sin(t);

a1=a*ct^2+b*ct*st+c*st^2;
c1=a*st^2-b*ct*st+c*ct^2;
----------------------------------------
```

```
pdft.m
--------------------------------------
function y=pdft(td,ux,uy,sx,sy)
% pdft.m 11-03-95
%
% Spatial Error Analysis ToolBox Version 1.0, October 5, 1997
% Copyright 1997-1998 by David Y. Hsu All rights reserved.
% dhsu@littongcs.com
%
% td in degrees
  sx2=sx^2; sy2=sy^2;
  d2r=pi/180;
  t=td*d2r;
  ct=cos(t);
  st=sin(t);
  ct2=ct^2;
  st2=st^2;
  A=ct2/sx2+st2/sy2;
  B=-2*(ux*ct/sx2+uy*st/sy2);
  C=(ux/sx)^2+(uy/sy)^2;
  w=B/sqrt(8*A);
  w2=w^2;
  y=1/(2*pi*sx*sy*A)*exp(-C/2)*(1-w*sqrt(pi)*erfc(w)*exp(w2));
--------------------------------------
```

E.5 M-FILES USED IN CHAPTER 5

```
nf2.m (see M-files used in Chapter 1)
----------------------------------------

cf2.m (see M-files used in Chapter 4)
----------------------------------------

gf2.m (see M-files used in Chapter 4)
----------------------------------------

sf1.m
----------------------------------------
function y=sf1(r)
% sf1.m  3-06-96
%
% Spatial Error Analysis ToolBox Version 1.0, October 5, 1997
% Copyright 1997-1998 by David Y. Hsu All rights reserved.
% dhsu@littongcs.com
```

```
%
% define f(r) = Maxwell Probability Density Function sigma = 1
y=sqrt(2/pi)*r.^2 .* exp(-r.^2/2);
-------------------------------------------
```

```
sf2.m
-------------------------------------------
function p=sf2(r)
% sf2.m 3-06-96
%
% Spatial Error Analysis ToolBox Version 1.0, October 5, 1997
% Copyright 1997-1998 by David Y. Hsu All rights reserved.
% dhsu@littongcs.com
%
% find p = F(r)
% where  F(r) = {integral of Maxwell pdf from 0 to r}, in terms of erf(x)

if r<0
  disp( 'input error ')
  disp( '  r>=0    ')
else
fac1 = sqrt(2/pi);
p = fac1*(-r.*exp(-r.^2/2))+erf(r/sqrt(2));
end
-------------------------------------------
```

```
sf3.m
-------------------------------------------
function r=sf3(p)
% sf3.m  2-19-96
%
% Spatial Error Analysis ToolBox Version 1.0, October 5, 1997
% Copyright 1997-1998 by David Y. Hsu All rights reserved.
% dhsu@littongcs.com
%
% given p and f(x), find r such that F(r) - p = 0,
% where F(r) = integral of f(x) from 0 to r.
%     Maxwell pdf used for f(x) here.
%
if p>1 | p<0
  disp( '0 <= p <= 1')
  error( 'input error for sf3(p)')
end
  if p==1
    r=Inf;
  else
    r=1;
    p0=sf2(r);
```

```
%%%%%%%% searching for a, b %%%%%%%%%%%%%%%
for j=1:2:5
 while (p0 > p)
  r=r-(0.1)^j ;
  p0=sf2(r);
 end
 while (p0 < p)
  r=r+(0.1)^j ;
  p0=sf2(r);
 end
end
end
-------------------------------------------

sf3a.m
-------------------------------------------
function y=sf3a(r)
% sf3a.m 2-19-96
%
% Spatial Error Analysis ToolBox Version 1.0, October 5, 1997
% Copyright 1997-1998 by David Y. Hsu All rights reserved.
% dhsu@littongcs.com
%
% given p, find G(r) = 0
% define G(r) = F(r) - p = {integral of trinormal pdf over [0,r]} - p
global p
% y = sf2(r) - p;
fac1 = sqrt(2/pi);
y = fac1*(-r.*exp(-r.^2/2))+erf(r/sqrt(2)) - p ;
-------------------------------------------

sf4.m
-------------------------------------------
function r=sf4(p)
% sf4.m 2-19-96
%
% Spatial Error Analysis ToolBox Version 1.0, October 5, 1997
% Copyright 1997-1998 by David Y. Hsu All rights reserved.
% dhsu@littongcs.com
%
% solve for r in G(r) = 0
% where G(r) = F(r) - p = {integral of trinormal pdf over [0,r]} - p
clear global p
global p
if p>1 | p<0
  disp( ' error in input ')
  disp( ' 0 <= p <= 1 ')
else
```

```
  if p==1
    r=Inf;
  elseif p==0
    r=0;
% else if p==0.025 | p==0.05
  else if p<=0.05 & p>0
%   r = fsolve('sf3a',0.5);
    r = newton(0.5,'sf3a','sf1',0.0001);
  else
%   r = fsolve('sf3a',1);
    r = newton(1,'sf3a','sf1',0.0001);
  end
end
------------------------------------------

newton.m (see M-files used in Chapter 1)
------------------------------------------

tf2.m
------------------------------------------
function p=tf2(r,u,v)
% tf2.m  2-20-96
%
% Spatial Error Analysis ToolBox Version 1.0, October 5, 1997
% Copyright 1997-1998 by David Y. Hsu  All rights reserved.
% dhsu@littongcs.com
%
% for triple integration over [0,r], u, v as parameters
% avoid u = 0 and v = 0  in tf1.m

c0=-1/2;
c1=(v^2-1)/(2*v^2);
c2=(u^2-1)/(2*u^2);
fac1=4/(sqrt(2*pi)*(pi*u*v));
mr2=32;   % 32
mw2=64 ;  % 64
mt2=36;   % 36

dr=r/mr2;
dw=1/mw2;
dt=(pi/2)/mt2;

%        w.r.t. r
for ir=1:mr2   % skip ir=0
r=dr*ir;
%        w.r.t. w
     for iw=0:mw2
     w=dw*iw;
```

```
%       w.r.t. t
        if c2==0 | w==1
         sum3=pi/2;
        else
         it=0:mt2;
         t=dt*it;
         y3=exp(c2*r.^2.*(1-w.^2).*cos(t).^2);
         sum3=sum(y3(1:mt2:mt2+1))+2*sum(y3(3:2:mt2-
                                           1))+4*sum(y3(2:2:mt2));

         sum3=sum3*dt/3;
        end
      if c1==0 | w==0
      y2(iw+1)=sum3;
      else
      y2(iw+1)=exp(c1*r.^2.*w.^2)*sum3 ;
      end
      end

sum2=sum(y2(1:mw2:mw2+1))+2*sum(y2(3:2:mw2-1))+4*sum(y2(2:2:mw2));
sum2=sum2*dw/3;
y1(ir+1)=exp(c0*r.^2).*r.^2*sum2;
end
sum1=y1(mr2+1)+2*sum(y1(3:2:mr2-1))+4*sum(y1(2:2:mr2));
p=fac1*sum1*dr/3;
if p>1
 p=1;
end
------------------------------------------

tf3.m
------------------------------------------
function r=tf3(p,u,v)
% tf3.m, 2-20-96
%
% Spatial Error Analysis ToolBox Version 1.0, October 5, 1997
% Copyright 1997-1998 by David Y. Hsu All rights reserved.
% dhsu@littongcs.com
%
% solve for r from F(r,u,v) - p = 0,
% sig(x)~ = sig(y)~ = sig(z)
if p>1 | p<0  | u<0 | v<0 | u>1 | v>1 | v>u
 error( 'input error for tf3(p)')
end
if p==1
  r=Inf;
else
  r=1;
  p0=tf2(r,u,v);
```

```
%%%%%%%% searching for a, b %%%%%%%%%%%%%%%%
 for j=1:2:5
   while (p0 > p)
     r=r-(0.1)^j;
     p0=tf2(r,u,v);
   end
   while (p0 < p)
     r=r+(0.1)^j;
     p0=tf2(r,u,v);
   end
 end
end
----------------------------------------
```

r2p3d.m

```
----------------------------------------
function r2p3d(k)
% r2p3d.m  2-26-96
%
% Spatial Error Analysis ToolBox Version 1.0, October 5, 1997
% Copyright 1997-1998 by David Y. Hsu All rights reserved.
% dhsu@littongcs.com
%
%
% for 3-dim normal distribution f(x,y,z) with sigma(x) = sigma(y) =
                                                          sigma(z)
% given k = multiple of SIGMA
% given k = multiple of MRE
% given k = multiple of RMSR
% given k = multiple of SEP
% r = k * SIGMA
% r = k * MSRE        (MRE = Mean Spherical Radial Error)
% r = k * RMSR        (RMSR = Root Mean Square Spherical Radial Error)
% r = k * SEP         (SEP = Spherical Error Probable)
% find p = F(r)
% where F(r) = integral of f(x) from 0 to r.
%    Maxwell pdf used for f(x) here.
%

 disp(' ')
 disp(' ')
 ic=menu('Options:','k*SIGMA','k*MRE','k*RMSR','k*SEP','all the
      above');

 if ic==1
 r= k;
 p=sf2(r) ;
 disp(['For r=' ,num2str(k), ' * SIGMA'])
```

```
disp(['p=', num2str(p)])
end

if ic==2
r= k* 2*sqrt(2/pi);
p=sf2(r)    ;
disp(['For r=' ,num2str(k), ' * MRE'])
disp(['p=', num2str(p)])
end

if ic==3
r= k*sqrt(3);
p=sf2(r)    ;
disp(['For r=' ,num2str(k), ' * RMSR'])
disp(['p=', num2str(p)])
end

if ic==4
r=k*sf4(0.5);
p=sf2(r) ;
disp(['For r=' ,num2str(k), ' * SEP'])
disp(['p=', num2str(p)])
end

if ic==5
r= k;
p=sf2(r)    ;
disp(['For r=' ,num2str(k), ' * SIGMA'])
disp(['   p=', num2str(p)])

r= k* 2*sqrt(2/pi);
p=sf2(r)      ;
disp(['For r=' ,num2str(k), ' * MRE'])
disp(['   p=', num2str(p)])

r= k*sqrt(3);
p=sf2(r)    ;
disp(['For r=' ,num2str(k), ' * RMSR'])
disp(['   p=', num2str(p)])

r=k*sf4(0.5);
p=sf2(r)     ;
disp(['For r=' ,num2str(k), ' * SEP'])
disp(['   p=', num2str(p)])
end
------------------------------------------
```

p2r3d.m

```
---------------------------------------------
function p2r3d(p)
% p2r3d.m  2-26-96
%
% Spatial Error Analysis ToolBox Version 1.0, October 5, 1997
% Copyright 1997-1998 by David Y. Hsu  All rights reserved.
% dhsu@littongcs.com
%
% r = r3ng(p)
% given p and f(x), find r such that F(r) - p = 0,
% express r in terms of SIGMA, MSRE, SEP, RMSR
% where F(r) = integral of f(x) from 0 to r.
%     Maxwell pdf used for f(x) here.
%
%

disp('  ')
disp('  ')
disp([' For p=', num2str(p) ])
ic=menu('options:', 'k*SIGMA', 'k*MRE', 'k*RMSR','k*SEP','all 4
     above');
r=sf4(p);

if ic==1
disp(['r=', num2str(r), ' times SIGMA'])
end

if ic==2
r=r/sqrt(8/pi);
disp(['r=', num2str(r), ' times MRE'])
end

if ic==3
r=r/sqrt(3);
disp(['r=', num2str(r), ' times RMSR'])
end

if ic==4
r=r/sf4(0.5);
disp(['r=', num2str(r), ' times SEP'])
end

if ic==5
disp(['r=', num2str(r), ' times SIGMA'])

r1=r/sqrt(8/pi);
disp(['r=', num2str(r1), ' times MSRE'])
```

```
r2=r/sqrt(3);
disp(['r=', num2str(r2), ' times RMSR'])

r3=r/sf4(0.5);
disp(['r=', num2str(r3), ' times SEP'])
end
------------------------------------------
```

edbvud.m

```
------------------------------------------
function p=edbvud(abc,sxyz)
% edbvud.m  2-20-96
%
% Spatial Error Analysis ToolBox Version 1.0, October 5, 1997
% Copyright 1997-1998 by David Y. Hsu All rights reserved.
% dhsu@littongcs.com
%
% The probability of a
% 3D joint ellipsoidal normal pdf with (sx, sy, sz), rho = 0
% over a box volume:  |x| < a,   |y| < b,  |z| < c.
% abc = [ a, b, c]
% sxyz = [sx,sy,sz]
%
a=abc(1);
b=abc(2);
c=abc(3);

sx=sxyz(1);
sy=sxyz(2);
sz=sxyz(3);

if sx<=0 | sy<=0 | sz<=0 | a<0 | b<0 | c <0
  error('check signs of input argument')
end
p=nf2(a/sx)*nf2(b/sy)*nf2(c/sz);
------------------------------------------
```

edbvui.m

```
------------------------------------------
function [a,b,c]=edbvui(p,k1,k2,sxyz)
% edbvui.m  2-28-96
%
% Spatial Error Analysis ToolBox Version 1.0, October 5, 1997
% Copyright 1997-1998 by David Y. Hsu All rights reserved.
% dhsu@littongcs.com
%
% Given the probability  sx, sy, sz, k1 = a/b, k2 = b/c, p
% Find a, b
```

```
%
% 3D joint elliptical normal pdf with sx, sy, sz, rho = 0
% over a rectangular region:  -a < x < a ,  -b < y < b, -c < z < c.
%
b=max(sxyz);
a=k1*b;
c=b/k2;
abc=[a b c];
p0=edbvud(abc,sxyz);

%%%%%%%% searching for a, b ,c %%%%%%%%%%%%%%%
for j=1:1:5
 while (p0 > p)
   b=b-(0.1)^j ;
   a=k1*b     ;
   c=b/k2     ;
abc=[a b c];
p0=edbvud(abc,sxyz);
 end

 while (p0 < p)
   b=b+(0.1)^j ;
   a=k1*b     ;
   c=b/k2     ;
abc=[a b c];
p0=edbvud(abc,sxyz);
 end
end
------------------------------------------

edsv.m
------------------------------------------
function p=edsv(r,u,v)
% edsv.m  2-20-96
%
% Spatial Error Analysis ToolBox Version 1.0, October 5, 1997
% Copyright 1997-1998 by David Y. Hsu  All rights reserved.
% dhsu@littongcs.com
%
% Ellipsoidal PDF, uncorrelated, spherical volume
%     u >= v

 if r<0 | u < 0  | v > 1  | u < v
   error('check input arguments in edsv.m')
 end
 mlim =0.01;

 if u<mlim
```

```
  u=0; v=0;
 end

 if v<mlim
  v=0;
 end

 if (u==0 & v==0)
  p=nf2(r) ;
 elseif (u==1 & v==0)
  p=cf2(r) ;
 elseif (u<1 & v==0)
  p=gf2(r,u);
 elseif (u==1 & v==1)
  p=sf2(r) ;
 else
  p=tf2(r,u,v);
 end
 ------------------------------------------

 sdsvud.m
 ------------------------------------------
 function p=sdsvud(R,sig)
 % sdsvud.m 2-20-96
 %
 % Spatial Error Analysis ToolBox Version 1.0, October 5, 1997
 % Copyright 1997-1998 by David Y. Hsu All rights reserved.
 % dhsu@littongcs.com
 %
 % for 3-dim normal distribution with sigma(x) = sigma(y) = sigma(z) = sig
 % find p = F(r)
 % where F(r) = integral of f(x) from 0 to r.
 % Maxwell pdf used for f(x) here.
 %
 if sig<=0 | R<0
  error(' check signs of input argument for sdsvud.m')
 end
 r=R/sig;
 p=sf2(r);
 ------------------------------------------

 sdsvui.m
 ------------------------------------------
 function R=sdsvui(p,sx)
 % sdsvui.m 2-20-96
 %
 % Spatial Error Analysis ToolBox Version 1.0, October 5, 1997
 % Copyright 1997-1998 by David Y. Hsu All rights reserved.
```

```
% dhsu@littongcs.com
%
R=sx*sf3(p);
-----------------------------------------
```

edsvud.m

```
-----------------------------------------
function p=edsvud(R,sxyz)
% edsvud.m 2-20-96
%
% Spatial Error Analysis ToolBox Version 1.0, October 5, 1997
% Copyright 1997-1998 by David Y. Hsu All rights reserved.
% dhsu@littongcs.com
%
% Ellipsoidal PDF, uncorrelated, spherical volume
% sxyz = [sx,sy,sz]

if R<0 | min(sxyz) < 0 | sxyz==[0 0 0]
  error('check input arguments in edsvud.m')
end

sx=sxyz(1);
sy=sxyz(2);
sz=sxyz(3);

sig=sort(sxyz);
r=   R/sig(3);
u=sig(2)/sig(3);
v=sig(1)/sig(3);

p=edsv(r,u,v);
-----------------------------------------
```

edsvui.m

```
-----------------------------------------
function R=edsvui(p,sxyz)
% edsvui.m 2-20-96
%
% Spatial Error Analysis ToolBox Version 1.0, October 5, 1997
% Copyright 1997-1998 by David Y. Hsu All rights reserved.
% dhsu@littongcs.com
%
% Ellipsoidal PDF, uncorrelated, spherical volume
% sxyz = [sx,sy,sz]

if p<0 | p>1 | min(sxyz) < 0 | sxyz==[0 0 0]
  error('check input arguments in edsvud.m')
end
```

```
R=mean(sxyz)*sf3(p);
p0 = edsvud(R,sxyz);
%%%%%%%% searching for a, b %%%%%%%%%%%%%%
for j=1:1:5
 while (p0 > p)
  R=R-(0.1)^j   ;
p0 = edsvud(R,sxyz);
 end
 while (p0 < p)
  R=R+(0.1)^j   ;
p0 = edsvud(R,sxyz);
 end
end
--------------------------------------------
```

```
edsvcd.m
--------------------------------------------
function p=edsvcd(R,sxyz,rho)
% edsvcd.m 2-05-96
%
% Spatial Error Analysis ToolBox Version 1.0, October 5, 1997
% Copyright 1997-1998 by David Y. Hsu All rights reserved.
% dhsu@littongcs.com
%
% Ellipsoidal PDF, correlated, spherical volume
% sxyz = [sx, sy, sz]
% rho = [r12, r13, r23]

if R<0 | min(sxyz)<=0 | abs(max(rho)) > 1
  error('check input arguments in edsvcd.m')
end

sx=sxyz(1);
sy=sxyz(2);
sz=sxyz(3);

r12=rho(1);
r13=rho(2);
r23=rho(3);

f1=1/(1+2*r12*r13*r23-r12^2-r13^2-r23^2);

if rho==[0 0 0]
  p=edsvud(R,[sx,sy,sz]);
else
  A= f1*(1-r23^2)/sx^2  ;
  B= f1*(1-r13^2)/sy^2  ;
```

```
 C= f1*(1-r12^2)/sz^2 ;

 q12=-2*(r12-r13*r23)*f1/(sx*sy);
 q13=-2*(r13-r23*r12)*f1/(sx*sz);
 q23=-2*(r23-r12*r13)*f1/(sy*sz);

 Q=[A q12/2 q13/2; q12/2 B q23/2; q13/2 q23/2 C];

 [V,D]=eig(Q);
 s1=1/sqrt(D(1,1));
 s2=1/sqrt(D(2,2));
 s3=1/sqrt(D(3,3));
 p=edsvud(R,[s1,s2,s3]);
end
----------------------------------------
```

```
edsvci.m
----------------------------------------
function R=edsvci(p,sxyz,rho)
% edsvci.m 2-27-96
%
% Spatial Error Analysis ToolBox Version 1.0, October 5, 1997
% Copyright 1997-1998 by David Y. Hsu  All rights reserved.
% dhsu@littongcs.com
%
% Ellipsoidal PDF, correlated, spherical volume
% sxyz = [sx,sy,sz]
% rho = [rxy,ryz,rxz]

if p<0 | p>1 | min(sxyz) < 0 | sxyz==[0 0 0] | min(rho)<-1 | max(rho)>1
  error('check input arguments in edsvci.m')
end

R=mean(sxyz)*sf3(p);
p0 = edsvcd(R,sxyz,rho) ;
%%%%%%%% searching for a, b %%%%%%%%%%%%%%%
for j=1:1:5
 while (p0 > p)
   R=R-(0.1)^j        ;
p0 = edsvcd(R,sxyz,rho) ;
 end
 while (p0 < p)
   R=R+(0.1)^j        ;
p0 = edsvcd(R,sxyz,rho) ;
 end
end
----------------------------------------
```

edevud.m

```
-----------------------------------------
function p=edevud(abc,sxyz)
% edevud.m 2-21-96
%
% Spatial Error Analysis ToolBox Version 1.0, October 5, 1997
% Copyright 1997-1998 by David Y. Hsu All rights reserved.
% dhsu@littongcs.com
%
% Ellipsoidal PDF, uncorrelated, ellipsoidal volume
% abc = [ a, b, c]
% sxyz = [sx,sy,sz]

if min(abc)<=0 | min(sxyz)<=0
  error('check input arguments in edevud.m')
end
sxyz= sxyz./abc;
p=edsvud(1,sxyz);
-----------------------------------------
```

edevui.m

```
-----------------------------------------
function [a,b,c]=edevui(p,k1,k2,sxyz)
%edevui.m 2-20-96
%
% Spatial Error Analysis ToolBox Version 1.0, October 5, 1997
% Copyright 1997-1998 by David Y. Hsu All rights reserved.
% dhsu@littongcs.com
%
% ellipsoidal normal pdf, x,y,z are uncorrelated
% ellipsoidal volume of integration
%   x^2/a^2 + y^2/b^2 + z^2/c^2 = 1
%
% k1 = a/b, k2 = b/c
%
sx=sxyz(1);
sy=sxyz(2);
sz=sxyz(3);

if sx< 0 | sy< 0 | sz <0 | k1 <0 | k2 <0 |sxyz==[0 0 0]
 error('check input arguments in edevui.m')
end

b=sf3(p)*mean(sxyz);
a=k1*b;
c=b/k2;
abc=[a b c]        ;
p0=edevud(abc,sxyz) ;
```

```
%%%%%%%% searching for a, b %%%%%%%%%%%%%%%
for j=1:1:5
 while (p0 > p)
  b=b-(0.1)^j;
  a=k1*b;
  c=b/k2 ;
  abc=[a b c]    ;
p0=edevud(abc,sxyz) ;
 end

 while (p0 < p)
  b=b+(0.1)^j;
  a=k1*b ;
  c=b/k2;
  abc=[a b c]    ;
p0=edevud(abc,sxyz) ;
 end
end
-------------------------------------------

edevcd.m
-------------------------------------------
function p=edevcd(abc,sxyz,rho)
% edevcd.m 2-05-96
%
% Spatial Error Analysis ToolBox Version 1.0, October 5, 1997
% Copyright 1997-1998 by David Y. Hsu All rights reserved.
% dhsu@littongcs.com
%
% Ellipsoidal PDF, correlated, ellipsoidal volume
% abc = [a, b, c]
% sxyz = [sx, sy, sz]
% rho = [r12, r13, r23]

if min(abc)<=0 | min(sxyz)<=0 | max(abs(rho)) > 1
  error('check input arguments in edevcd.m')
end

if rho== [0 0 0]
  p=edevud(abc,sxyz);
else

a=abc(1);
b=abc(2);
c=abc(3);

sx=sxyz(1)/a;
```

```
sy=sxyz(2)/b;
sz=sxyz(3)/c;

r12=rho(1);
r13=rho(2);
r23=rho(3);

f1=1/((1-r12^2)*(1-r13^2)*(1-r23^2));

  A= f1*(1-r23^2)/sx^2 ;
  B= f1*(1-r13^2)/sy^2 ;
  C= f1*(1-r12^2)/sz^2 ;

  q12=-2*(r12-r13*r23)*f1/(sx*sy);
  q13=-2*(r13-r23*r12)*f1/(sx*sz);
  q23=-2*(r23-r12*r13)*f1/(sy*sz);
  Q=[A q12/2 q13/2; q12/2 B q23/2; q13/2 q23/2 C];

  [V,D]=eig(Q);
  s1=1/sqrt(D(1,1));
  s2=1/sqrt(D(2,2));
  s3=1/sqrt(D(3,3));
  p=edsvud(1,[s1,s2,s3]);

end
------------------------------------------

edevci.m
------------------------------------------
function [a,b,c]=edevci(p,k1,k2,sxyz,rho)
%edevci.m 2-20-96
%
% Spatial Error Analysis ToolBox Version 1.0, October 5, 1997
% Copyright 1997-1998 by David Y. Hsu  All rights reserved.
% dhsu@littongcs.com
%
% ellipsoidal normal pdf, x,y,z are correlated
% ellipsoidal volume of integration
%   x^2/a^2 + y^2/b^2 + z^2/c^2 = 1
%
% k1 = a/b, k2 = b/c
%
sx=sxyz(1);
sy=sxyz(2);
sz=sxyz(3);
z3=[0 0 0]

if sx< 0|sy< 0|sz <0|k1 <0|k2 <0|sxyz==z3|min(rho)<-1|max(rho)>1
```

```
error('check input arguments in edevci.m')
end

b=sf3(p);
a=k1*b;
c=b/k2;
abc=[a b c]
p0=edevcd(abc,sxyz,rho)
%%%%%%%% searching for a, b %%%%%%%%%%%%%%
for j=1:1:5
 while (p0 > p)
   b=b-(0.1)^j;
   a=k1*b;
   c=b/k2 ;
   abc=[a b c]
 p0=edevcd(abc,sxyz,rho)
  end
  while (p0 < p)
   b=b+(0.1)^j;
   a=k1*b ;
   c=b/k2;
   abc=[a b c]
 p0=edevcd(abc,sxyz,rho)
  end
 end
 ------------------------------------------
```

E.6 M-FILES USED IN CHAPTER 6

```
psi.m
------------------------------------------
function y=psi(x)
% psi.m 6-21-96
%
% Spatial Error Analysis ToolBox Version 1.0, October 5, 1997
% Copyright 1997-1998 by David Y. Hsu All rights reserved.
% dhsu@littongcs.com
%
global n

if x==1
y=n;
else
y=(1-x.^n)./(1-x); % use 1-eps to avoid singularity when integrating
end
------------------------------------------
```

psi1.m

```
-----------------------------------------
% psi1.m  3-12-97
%
% Spatial Error Analysis ToolBox Version 1.0, October 5, 1997
% Copyright 1997-1998 by David Y. Hsu All rights reserved.
% dhsu@littongcs.com
%
% GM/RMS vs N
% part I
clf
eu=.5772156649;
x=[(.1:0.025:.2) (0.3:.1:.9) (1:9) (10:10:100)]  ;
Nx = length(x);

for j=1:Nx
 n2=x(j);   m=n2/2;
  y=psinh(n2);
% y=psin(m);
 z(j)=exp(y/2)/sqrt(m);
end

DAT=[x' z']
save c:\mfile\psi.mat x z

fname='c:\mfile\psi.dat'
fid=fopen(fname,'wt')
fprintf(fid,'%8.4f %8.4f\n',DAT')
fclose(fid)
%%%%%%%%%%%%%%%%%%%%%%%%%%%%
% part II
load c:\mfile\psi.mat

semilogx(x,z, x,z,'o')
axis('square')
xlabel('n')
ylabel('GM/RMS')
% title(' GM/RMS vs. n ')
grid
figure(1)
pf3('f61')
xyz % stop on purpose
-----------------------------------------
```

psin.m

```
-----------------------------------------
function y=psin(x)
```

```
% psin.m 3-14-96
%
% Spatial Error Analysis ToolBox Version 1.0, October 5, 1997
% Copyright 1997-1998 by David Y. Hsu All rights reserved.
% dhsu@littongcs.com
%
% evaluate psi(x)
% psin(x) = psinh(2x)
% for x = integers, integer + 0.5, and 0<x<1
% recursive formula used for other values of x

eu=.5772156649;
global n
n=x;

 if n<0
  y=psin(1-x)-pi/tan(pi*x);
 else
 if n<1                        % n is less than 1
  y=quad8('psi',0,1-eps)-eu-1/n;    % 1 - eps to avoid singularity
% n: integer                    % n is integer
 elseif n-floor(n)==0 & n>=1
    k=1:n-1 ; s=1./k;
    y= -eu + sum(s) ;
% n: integer + 0.5                  % n is integer + 0.5
 elseif n-floor(n)==1/2 & n>1
    m=(2*n-1)/2;
    k=1:m ; s=1./(2*k-1);
    y= -eu -2*log(2) + 2*sum(s) ;
 else
    x=n-1;
    y=psin(x)+ 1/x;
 end
 end

 clear global n
 ------------------------------------------


 psinh.m
 ------------------------------------------
 function y=psinh(n2)
 % psinh.m 3-14-96
 %
 % Spatial Error Analysis ToolBox Version 1.0, October 5, 1997
 % Copyright 1997-1998 by David Y. Hsu All rights reserved.
 % dhsu@littongcs.com
 %
 % evaluate psi(n/2)
```

```
% psinh(2x) = psin(x)
% for n = integers and 0<n<1

eu=.5772156649;
 global n
 n=n2/2;
 m=floor(n);

% evaluate psi
 if n2<1 & n2 >0              % n2 is less than 1
  y=quad8('psi',0,1-eps)-eu-1/n;      % 1 - eps to avoid singularity
 end
% odd n2: 1,3,5,...,99
  if n2==m*2+1                % n2 is odd
   k=1:m ;  s=1./(2*k-1);
   y=-eu - 2*log(2) + 2*sum(s) ;
  end
% even n2: 2,4,6,...,100
  if n2==m*2                 % n2 is odd
   k=1:m-1;  s=1./k;           % n2 is even
   y=-eu + sum(s) ;
  end
clear global n
% z = GM/RMS
% z = exp(y/2)/sqrt(n);
-----------------------------------------

x2df.m (see M-files used in Chapter 2)
-----------------------------------------

x2dg.m (see M-files used in Chapter 2)
-----------------------------------------

x2cdf.m (see M-files used in Chapter 2)
-----------------------------------------

x2disc2.m
-----------------------------------------
function y=x2disc2(x)
% x2disc2.m  1-16-94
%
%  Spatial Error Analysis ToolBox Version 1.0, October 5, 1996
%  Copyright 1996-1997 by David Y. Hsu  All rights reserved.
%  dhsu@littongcs.com
%
% define G(a) = F(a) - p
% where F(a) = [integral of f(x) from 0 to a]
%     x - distribution of degree n + 2 is used for f(x) here.
```

```
%
global p n
% chi-square pdf f(t,n) is singular at t = 0 for n<2
% when integrating
% use F(x,n) = g(x,n) + F(x,n + 2) relationship for n<2
% g(x,n)= (x/2)^(n/2).*exp(-x/2)/gamma(n/2 + 1);
% F(x,n) = Integral of f(t,n) from 0 to x
% f(t,n) = chi-square pdf with degree n
% x2cdf.m for F(x,n) and g(x,n) + F(x,n + 2)

 y=x2cdf(x) - p;
------------------------------------------

chi1.m
------------------------------------------
% chi1.m 1-15-94
%
% Spatial Error Analysis ToolBox Version 1.0, October 5, 1997
% Copyright 1997-1998 by David Y. Hsu All rights reserved.
% dhsu@littongcs.com
%
clear
ic=2;  % skip step 1

%%%%%%%%%%%%%%%%%%%%%%%%%%%%%%%%%%%%
if ic==1
% part I
%      use y = sqrt(x/n), x has chi-square distribution with deg n
clear global n p
global n p

q=[0.5 .8 .9 .95];
l=[(0.1:0.025:0.2) (0.3:.1:.9) (1:9) (10:10:100)]';
NL = length(l);
NQ = length(q);

for jj=1:NQ
  p=q(jj);
  x =1;     %initial guess for iteration
% x =0.01;  %initial guess for fsolve
for ii=1:NL
  n=l(ii);
%%%%%%%%%%%%%%%%%%%%%%%%%%%%%%%%%%%%%%%%%%%
%% use fsolve from optimization toolbox %%
%% xs = fsolve('x2disc2',x);
%% xo = xs;
%%%%%%%%%%%%%%%%%%%%%%%%%%%%%%%%%%%%%%%%%%%
%%%%%% iteration %%%%%%%%%%%
```

```
      p0=x2cdf(x);
    for j=1:2:9
      while (p0 > p)
      x=x-(0.1)^j;
      p0=x2cdf(x);
      end

      while (p0 < p)
      x=x+(0.1)^j;
      p0=x2cdf(x);
      end
    end
     xo=x;
%%%%%%%%%%%%%%%%%%%%%%%%%%%%%%
 r(ii,jj)=sqrt(xo/n)
 pause
end
end

DAT =[l  r];
save \mfile\chi.mat  l r

fname='c:\mfile\chi.dat'
fid=fopen(fname,'wt')
fprintf(fid,'%8.2f %8.4f %8.4f %8.4f %8.4f\n',DAT')
fclose(fid)
end
%%%%%%%%%%%%%%%%%%%%%%%%%%%%%%%%%%%%%
if ic==2
% part II
load c:\mfile\chi.mat
NQ =4
adj= [ 0.7  0.5  0.15  -0.1];

semilogx(l,r)
xlabel('n')
ylabel(' R(p)/RMS ')
% title(' R(p)/RMS vs n,  p=0.5, 0.8, 0.9, 0.95')
grid

for kk=1:NQ
if kk==1  p=0.5
  elseif kk==2 p=0.8
  elseif kk==3 p=0.9
  elseif kk==4 p=0.95
end

  text(.5,r(5,kk)+adj(kk),['p=' num2str(100*p) '%'])
```

```
end
axis('square')
figure(1)
pause
pf3('f62')
end
------------------------------------------

chi2.m
------------------------------------------
% chi2.m  10-1-97
%
% Spatial Error Analysis ToolBox Version 1.0, October 5, 1997
% Copyright 1997-1998 by David Y. Hsu  All rights reserved.
% dhsu@littongcs.com
%
% R(p)/RMS=r  vs  GM/RMS=z

clear; clf;

% result from psi1.m
load c:\mfile\psi.dat
z=psi(:,2);

% result from chi1.m
load c:\mfile\chi.dat
r=chi(:,2:5);

DAT=[z r]
save c:\mfile\psichi.mat z r


fname='c:\mfile\psichi.dat'
fid=fopen(fname,'wt')
fprintf(fid,'%8.4f%8.4f%8.4f%8.4f%8.4f\n',DAT')
fclose(fid)

plot(z,r)
grid

adj = [0 0 0.06 0.02];
for kk=1:4
if kk==1  p=0.5    ;
  elseif kk==2 p=0.8 ;
  elseif kk==3 p=0.9 ;
  elseif kk==4 p=0.95;
end
```

```
 text(0.2,r(7,kk)+adj(kk),['p=' num2str(100*p) '%'])
end

xlabel(' x=GM/RMS ')
ylabel(' y=R(p)/RMS ')
% title(' R(p)/RMS vs z=GM/RMS, for p=0.5, 0.8, 0.9, 0.95')
axis('square')
figure(1)
pf3('f63')
------------------------------------------

gmrms2n.m
------------------------------------------
function n=gmrms2n(gm,rms)
% gmrms2n.m 3-14-96
%
% Spatial Error Analysis ToolBox Version 1.0, October 5, 1997
% Copyright 1997-1998 by David Y. Hsu All rights reserved.
% dhsu@littongcs.com
%
fid=fopen('c:\mfile\psi.dat','r');
dat=fscanf(fid,'%g %g',[2,inf]);
fclose(fid);
ratio=gm/rms ;

n2=dat(1,:);
r2=dat(2,:);
n =interp1(r2,n2,ratio,'spline');
------------------------------------------

n2gmrms.m
------------------------------------------
function ratio=n2gmrms(n)
% n2gmrms.m 3-14-96
%
% Spatial Error Analysis ToolBox Version 1.0, October 5, 1997
% Copyright 1997-1998 by David Y. Hsu All rights reserved.
% dhsu@littongcs.com
%
y=psin(n/2);
ratio=exp(y/2)/sqrt(n/2);
------------------------------------------

r2pc6.m
------------------------------------------
function p=r2pc6(gm,rms,r)
% r2pc6.m 6-22-96
%
```

```
% Spatial Error Analysis ToolBox Version 1.0,  October 5, 1996
% Copyright 1996-1997 by David Y. Hsu All rights reserved.
% dhsu@littongcs.com
%
% when n = 0.9358  and r = 14.7497
% gm=5.81, rms=11.53          ---> p=0.8000.
clear global n
global n

fid=fopen('c:\mfile\psi.dat','r');
dat=fscanf(fid,'%g %g',[2,inf]);
fclose(fid);

ratio=gm/rms;

n2=dat(1,:);
r2=dat(2,:);

n =interp1(r2,n2,ratio,'spline');
ms=rms^2;
x=n*r^2/ms;

   p=x2cdf(x) ;
-------------------------------------------


p2rc6.m
-------------------------------------------
function R=p2rc6(gm,rms,p)
% p2rc6.m  1-2-96
%
% Spatial Error Analysis ToolBox Version 1.0, October 5, 1997
% Copyright 1997-1998 by David Y. Hsu All rights reserved.
% dhsu@littongcs.com
%
% when n = 0.9358
% gm = 5.8054,  rms = 11.5296,  p = 0.5 ---> r = 7.542
clear global n
global n

fid=fopen('c:\mfile\psi.dat','r');
dat=fscanf(fid,'%g %g',[2,inf]);
fclose(fid);

ratio=gm/rms;
n2=dat(1,:);
r2=dat(2,:);
n =interp1(r2,n2,ratio,'spline');
```

```
%%%%%% use fsolve.m (from optimization toolbox) %%%%%%
% global p
% x = fsolve('x2disc2',1) ;
%%%%%%%%%%%%%%%%%%%%%%%%%%%%%%%%%%%%%%%%%%%%%%%%%%%%%%%%%%

%%%%%% iteration %%%%%%%%%%%
if p==1
  x=Inf;
else
  x=1;  % initial guess
    p0= x2cdf(x);
  for j=1:2:5
    while (p0 > p)
    x=x-(0.1)^j;
    p0= x2cdf(x);
    end

    while (p0 < p)
    x=x+(0.1)^j;
    p0= x2cdf(x);
    end
 end
end
%%%%%%%%%%%%%%%%%%%%%%%%%%%%%

  disp([' n = ', num2str(n)])
  R=rms*sqrt(x/n);
------------------------------------------

gmrms.m
------------------------------------------
function [gm , rms, ratio]=gmrms(fname)
% gmrms.m  6-25-96
%
% Spatial Error Analysis ToolBox Version 1.0, October 5, 1997
% Copyright 1997-1998 by David Y. Hsu All rights reserved.
% dhsu@littongcs.com
%
xo=myfun1(fname)

[m,n]=size(xo);
if n==1
 x=xo ;
end

if n==2
 x=xo(:,1).^2 + xo(:,2).^2 ;
 x=sqrt(x);
end
```

```
if n==3
 x=xo(:,1).^2 + xo(:,2).^2  + xo(:,3).^2;
 x=sqrt(x);
end

 tmp=prod(x);
 gm=tmp^(1/m);

 x2=x.^2;
 tmp2=sum(x2);
 rms=sqrt(tmp2/m);

 ratio=gm/rms;
-------------------------------------------
```

```
myfun1.m
-------------------------------------------
function x=myfun1(fname)
% myfun1.m  2-14-96
%
% Spatial Error Analysis ToolBox Version 1.0, October 5, 1997
% Copyright 1997-1998 by David Y. Hsu  All rights reserved.
% dhsu@littongcs.com
%
  s1=['load ' 'c:\mfile\' fname]
  eval(s1)
  eval(['x = ', fname(1:3) ';'])
-------------------------------------------
```

E.7 M-FILES USED IN CHAPTER 7

```
bf1.m
-------------------------------------------
function z=bf1(r)
% bf1.m   r>=0,    1-15-96
%
% Spatial Error Analysis ToolBox Version 1.0, October 5, 1997
% Copyright 1997-1998 by David Y. Hsu  All rights reserved.
% dhsu@littongcs.com
%
% define f(r) = 2-dim normal pdf, rho = 0, sigx and sigy may be unequal.
global mu

k1= r.^2*(1+mu^2)/(4*mu^2)        ;
k2= r.^2*(1-mu^2)/(4*mu^2)        ;
```

```
z= r/mu.* exp(-k1).*besseli(0,k2) ;
-------------------------------------------

bf2.m
-------------------------------------------
function y=bf2(r,mu)
% bf2.m  1-15-96
%
% Spatial Error Analysis ToolBox Version 1.0, October 5, 1997
% Copyright 1997-1998 by David Y. Hsu All rights reserved.
% dhsu@littongcs.com
%
% Find probability  p
% function p = bf2(r,mu)
% where F(r) = integral of f(r,mu) over [0,r]
clear global mu
global mu

if r<0 | mu>1 | mu<0
     disp( ' input error for bf2(r,mu)' )
     disp( '   r >= 0 ' )
     disp( ' 0<= mu <= 1' )

elseif mu==0
     y=nf2(r);

elseif mu==1
     y=cf2(r);

else
     y=quad('bf1',0,r);
end
-------------------------------------------

bf3.m
-------------------------------------------
function y=bf3(r)
% bf3.m  2-14-96
% p, mu are given parameters
%
% Spatial Error Analysis ToolBox Version 1.0, October 5, 1997
% Copyright 1997-1998 by David Y. Hsu All rights reserved.
% dhsu@littongcs.com
%
% define G(r) = F(r) - p = {integral of binormal pdf over [0,r]} - p
global p mu
y=quad('bf1',0,r) - p;
-------------------------------------------
```

```
bf4.m
------------------------------------------
function r=bf4(p,mu)
% bf4.m 2-14-94
%
% Spatial Error Analysis ToolBox Version 1.0, October 5, 1997
% Copyright 1997-1998 by David Y. Hsu All rights reserved.
% dhsu@littongcs.com
%
% find r for G(r) = F(r) - p = 0

clear global p mu
global p mu

if p>1 | p<0 | mu>1 | mu < 0
  disp( ' input error for bf4(p,mu) ')
  disp( ' 0 <= p  <= 1 ')
  disp( ' 0 <= mu <= 1 ')

elseif mu==1
  r=cf3(p);

elseif mu==0
  r=nf3(p);

else
  if p==1
    r=Inf;
  else
t1=clock;
%   r = fsolve('bf3',1);
    r=newton(1,'bf3','bf1',0.0001);
time_used=etime(clock,t1)
  end
end
------------------------------------------

cf2.m  (see M-files used in Chapter 4)
------------------------------------------

cf3.m  (see M-files used in Chapter 4)
------------------------------------------

ang2p.m
------------------------------------------
function p=ang2p(ang,p0)
% ang2p.m
```

```
%
% Spatial Error Analysis ToolBox Version 1.0, October 5, 1997
% Copyright 1997-1998 by David Y. Hsu All rights reserved.
% dhsu@littongcs.com
%
ro=sqrt(-2*log(1-p0))
d2r=pi/180;

   ang=ang*d2r;
   s1=csc(ang/2)/sqrt(2);
   s2=sec(ang/2)/sqrt(2);
   sx=max(s1,s2)
   sy=min(s1,s2);
   u=sy/sx
   mu=u;
   r=ro/sx
   p=bf2(r,mu)*100;
------------------------------------------

fsca.m
------------------------------------------
function [sf, af, fac]=fsca(sa,sb,ang)
% fsca.m  3-11-96
%
% Spatial Error Analysis ToolBox Version 1.0, October 5, 1997
% Copyright 1997-1998 by David Y. Hsu All rights reserved.
% dhsu@littongcs.com
%
% p. 1219 Bowditch, Fig. 6qi
% Find fictitious sigma, sigma factor, and fictitious cut angle
% ang, af in degrees

d2r=pi/180;
u=min(sa,sb)/max(sa,sb);

b=atan(u);
 sf=sin(2*b).*sqrt(sa^2+sb^2)/sqrt(2);
 af=asin(sin(2*b)*sin(ang*d2r))/d2r;
 fac = sf/max(sa,sb);

 disp([' fictitious sigma = ' num2str( sf )])
 disp([' fictitious cut angle = ' num2str(af) ' degrees'] )
 disp([' sigma factor = ' num2str(fac)] )
------------------------------------------

sem1.m
------------------------------------------
function p=sem1(r,s1,s2,ang)
```

```
% sem1.m 4-18-96
%
% Spatial Error Analysis ToolBox Version 1.0, October 5, 1997
% Copyright 1997-1998 by David Y. Hsu All rights reserved.
% dhsu@littongcs.com
%
% from Burt, single ellipse method #1
% ang = cut angle in degrees

d2r=pi/180 ;
% fictitious sigma and fictitious cut angle
  [sf, af, fac]=fsca(s1,s2,ang);

sa= sf/sqrt(2)/sin(af*d2r/2) ;
sb= sf/sqrt(2)/cos(af*d2r/2) ;
p=edcaud(r,sa,sb);
-------------------------------------------

edcaud.m (see M-files used in Chapter 4)
-------------------------------------------

sem2.m
-------------------------------------------
function p=sem2(r,sa,sb,ang)
% sem2.m 4-18-96
%
% Spatial Error Analysis ToolBox Version 1.0, October 5, 1997
% Copyright 1997-1998 by David Y. Hsu All rights reserved.
% dhsu@littongcs.com
%
% Example from Burt  method 2
% given r, s1, s2 and cut angle find probability
% p = P(r,s1,s2,cut_angle)

d2r=pi/180;
sa2=sa^2;
sb2=sb^2;
f=sin(ang*d2r)^2;

sx2=(sa2+sb2+sqrt((sa2+sb2)^2-4*f*sa2*sb2))/(2*f);
sy2=(sa2+sb2-sqrt((sa2+sb2)^2-4*f*sa2*sb2))/(2*f);

sx=sqrt(sx2)                    ;
sy=sqrt(sy2)                    ;
mu=min(sx,sy)/max(sx,sy) ;
k=r/max(sx,sy);
p=bf2(k,mu)  ;
clc
```

```
disp(['P(r)= ' num2str(p*100) '%, for r= ' num2str(r)])
disp(['r= ' num2str(k) ' times sigma(x)' ])
disp(['sigma(x)= ' num2str(max(sx,sy)) ])
disp(['sigma(y)= ' num2str(min(sx,sy)) ])
disp(['u= ' num2str(mu) ])
------------------------------------------

noval.m
------------------------------------------
function p=noval(r)
% noval.m  3-11-96
%
% Spatial Error Analysis ToolBox Version 1.0, October 5, 1997
% Copyright 1997-1998 by David Y. Hsu All rights reserved.
% dhsu@littongcs.com
%
% multiple ellipses all axes parallel
% r = input ('r = (20 in p.32 Burt) ')            ;
load c:\mfile\nsd.dat

x=nsd(:,1)
y=nsd(:,2)

sa=sqrt(sum(x.*x))
sb=sqrt(sum(y.*y))

ang=90;
% ang=input ('cut angle=  (90 in p.32 Burt) ')  ;

disp(' ')
d2r=pi/180;
sa2=sa^2;
sb2=sb^2;
f=sin(ang*d2r)^2;

sx2=(sa2+sb2+sqrt((sa2+sb2)^2-4*f*sa2*sb2))/(2*f);
sy2=(sa2+sb2-sqrt((sa2+sb2)^2-4*f*sa2*sb2))/(2*f);

sx=sqrt(sx2)                      ;
sy=sqrt(sy2)                      ;
mu=min(sx,sy)/max(sx,sy) ;
k=r/max(sx,sy);
p=bf2(k,mu)   ;
clc
disp(['p(r)= ' num2str(p*100) '%, for r= ' num2str(r)])
disp(['r= ' num2str(k) ' times sigma(x)' ])
disp(['sigma(x)= ' num2str(max(sx,sy)) ])
disp(['sigma(y)= ' num2str(min(sx,sy)) ])
```

```
disp(['u= ' num2str(mu) ])
------------------------------------------

noval1.m
------------------------------------------
function R=noval1(p)
% noval1.m  3-11-96
%
% Spatial Error Analysis ToolBox Version 1.0, October 5, 1997
% Copyright 1997-1998 by David Y. Hsu  All rights reserved.
% dhsu@littongcs.com
%
% multiple ellipses with random axes orientations
% p = input( 'p =  (.5 in p.40 Burt) ');

d2r=pi/180;

load c:\mfile\nsd1.dat

nsd1
x=nsd1(:,1)      ;
y=nsd1(:,2)      ;
th=nsd1(:,3)*d2r  ;

a=cos(th).^2./x.^2 + sin(th).^2./y.^2
b=cos(th).^2./y.^2 + sin(th).^2./x.^2
c=cos(th).*sin(th).*(1./y.^2 - 1./x.^2)

rho=c./sqrt(a.*b)
w2=1./((1-rho.^2).*a)
z2=1./((1-rho.^2).*b)

sw2f=sum(w2)
sz2f=sum(z2)
rhof=sum(rho .*sqrt(w2) .*sqrt(z2))/(sqrt(sw2f)*sqrt(sz2f))

sx2f=0.5*(sw2f+sz2f+sqrt((sw2f+sz2f)^2-4*sw2f*sz2f*(1-rhof^2)))
sy2f=0.5*(sw2f+sz2f-sqrt((sw2f+sz2f)^2-4*sw2f*sz2f*(1-rhof^2)))

thf=atan2(2*rhof*sqrt(sw2f*sz2f),sw2f-sz2f)/d2r/2

sa=sqrt(sx2f)
sb=sqrt(sy2f)
% -----------------------------------------------

mu=min(sa,sb)/max(sa,sb) ;

K=bf4(p,mu)           ;
```

```
R=K*max(sa,sb)           ;

disp(['R=CEP(' num2str(p*100) ')= ' num2str(R)])
disp(['R= ' num2str(K) ' times max{ sigma(x),sigma(y) }' ])
disp(['sigma(x)= ' num2str(max(sa,sb)) ])
disp(['sigma(y)= ' num2str(min(sa,sb)) ])
disp(['u= ' num2str(mu) ])
------------------------------------------
```

E.8 M-FILES USED IN CHAPTER 8

```
cdfchi2.m
------------------------------------------
function p=cdfchi2(x,N)
% cdfchi2.m  10-01-97
%
% Spatial Error Analysis ToolBox Version 1.0, October 5, 1997
% Copyright 1997-1998 by David Y. Hsu All rights reserved.
% dhsu@littongcs.com
%

  s=['c:\mfile\p2x' num2str(N)];
  s1=['load ' s];
  eval(s1)
  p = interp1(xx,pp,x)
------------------------------------------

ichi2.m
------------------------------------------
function x=ichi2(p,N)
% ichi2.m 10-01-97
%
% Spatial Error Analysis ToolBox Version 1.0, October 5, 1997
% Copyright 1997-1998 by David Y. Hsu All rights reserved.
% dhsu@littongcs.com
%
  s=['c:\mfile\p2x' num2str(N)];
  s1=['load ' s];
  eval(s1)
  x = interp1(pp,xx,p);
------------------------------------------

an2ka.m
------------------------------------------
function Ka=an2ka(a,N)
% an2ka.m 9-29-97
```

```
%
% Spatial Error Analysis ToolBox Version 1.0, October 5, 1996
% Copyright 1996-1997 by David Y. Hsu All rights reserved.
% dhsu@littongcs.com
%
% given a and N,  find Ka

% if statistics toolbox is available, use the command chi2inv.m
%  ta = chi2inv(1 - a,2 * N - 1);

% if statistics toolbox is not available, to replace chi2inv.m
   ta = ichi2(1 - a,2 * N - 1);

Ka=sqrt(ta/(2*N));
----------------------------------------

anl2b.m
----------------------------------------
function b=anl2b(a,N,lamb)
% anl2b.m  9-29-97
%
% Spatial Error Analysis ToolBox Version 1.0, October 5, 1996
% Copyright 1996-1997 by David Y. Hsu All rights reserved.
% dhsu@littongcs.com
%
% given a, N and lambda, find b

% if statistics toolbox is available
% ta = chi2inv(1 - a,2 * N - 1);

% if statistics toolbox is not available, replace chi2inv with ichi2
  ta = ichi2(1 - a,2 * N - 1);

Ka=sqrt(ta./(2 * N))
tb=ta/lamb^2;

% if statistics toolbox is available
% b = chi2cdf(tb,2 * N - 1);

% if statistics toolbox is not available, replace chi2cdf with cdfchi2
  b = cdfchi2(tb,2 * N - 1);
----------------------------------------

cln2a.m
----------------------------------------
function a=cln2a(C,L,N)
% cln2a.m  9-29-97
%
```

```
% Spatial Error Analysis ToolBox Version 1.0, October 5, 1996
% Copyright 1996-1997 by David Y. Hsu  All rights reserved.
% dhsu@littongcs.com
%
   Ka=L/C;
   ta=2*N*Ka^2;
% if statistics toolbox is available
%   a = (1 - chi2cdf(ta,2 * N - 1)) * 100 ;

% if statistics toolbox is not available, replace chi2cdf with cdfchi2
   a = (1 - cdfchi2(ta,2 * N - 1)) * 100 ;
------------------------------------------
```

```
cln2b.m
------------------------------------------
function b=cln2b(C,L,N,lamb)
% cln2b.m  10-01-97
%
% Spatial Error Analysis ToolBox Version 1.0, October 5, 1996
% Copyright 1996-1997 by David Y. Hsu  All rights reserved.
% dhsu@littongcs.com
%
   Ka=L/C;
   ta=2*N*Ka^2;
% if statistics toolbox is available
%   a = (1 - chi2cdf(ta,2 * N - 1)) * 100;

% if statistics toolbox is not available, replace chi2cdf with cdfchi2
   a = (1 - cdfchi2(ta,2 * N - 1)) * 100;

   tb=2*N*(Ka./lamb).^2;
% if statistics toolbox is available
%   b = chi2cdf(tb,2 * N - 1) * 100   ;

% if statistics toolbox is not available, replace chi2cdf with cdfchi2
   b = cdfchi2(tb,2 * N - 1) * 100   ;
------------------------------------------
```

Bibliography

[1] H. J. Larson, *Introduction to Probability and Statistical Inference*, John Wiley, 1974.

[2] P. Hoel, *Introduction to Mathematical Statistics*, 4th edition, John Wiley, 1971.

[3] R. Burton and J. Faires, *Numerical Analysis*, PWS-KENT Publishing Company, 1989.

[4] P. Meyer, *Introductory Probability and Statistical Applications*, 2nd edition, Addison-Wesley, 1970.

[5] J. E. Freund, *Mathematical Statistics*, 2nd edition, Prentice Hall, 1971.

[6] P. Bevington and K. Robinson, *Data Reduction and Error Analysis for the Physical Sciences*, 2nd edition, McGraw-Hill, 1992.

[7] A. W. Burt, D. J. Kaplan, R. R. Keenly, J. F. Reeves, and F. B. Shaffer, *Mathematical Considerations Pertaining to the Accuracy of Position Location and Navigation Systems*, Part I, Stanford Research Institute, Menlo Park, CA, Research Memorandum NWRC-RM34, NTIS #AD629-609, 1966.

[8] N. Bowditch, *American Practical Navigator*, Volume 1, DMAHC, 1977.

[9] B. P. Blair, *Statistical Procedures for Analysis of Weapon System Accuracy*, Autonetics EM-1188, 1959.

[10] G. R. Pitman, Jr., *Inertial Guidance*, John Wiley, 1962.

[11] F. Mason, *Maximum Likelihood Estimation of the Distribution of Radial Errors*, AD-A950-316, 1965.

[12] M. Abramowitz and I. Stegun, *Handbook of Mathematical Functions*, U.S. Department of Commerce, National Bureau of Standards, 1964.

[13] AIR STD 53/16, *The Specification for Evaluation of the Accuracy of Hybrid Navigation Systems*, 1984.

[14] R. Kalafus and G. Chin, *Measures of Accuracy in the NAVSTAR/GPS: 2DRMS VS. CEP*, ION, 1984.

[15] A. Papoulis, *Probability and Statistics*, Prentice Hall, 1990.

[16] *Operating Characteristic Curves Associated with Pass/Fail Criteria of CAINS II and SAHRS Acceptance Test Procedures*, SofTech, 1987.

[17] A. Leick, *GPS Satellite Surveying*, Wiley International, 1990.

[18] G. Strang and K. Borre, *Linear Algebra, Geodesy and GPS*, Wellesley Cambridge Press, 1997.

[19] D. Y. Hsu, *Relation between DOPs and volume of Tetrahedron formed by 4 Satellites*, IEEE 1994 PLANS Symposium.

Index